T0093890

An Analysis of

Richard Dawkins's

The Selfish Gene

Nicola Davis

Published by Macat International Ltd
24:13 Coda Centre, 189 Munster Road, London SW6 6AW.

Distributed exclusively by Routledge
2 Park Square, Milton Park, Abingdon, Oxon OX14 4RN
711 Third Avenue, New York, NY 10017, USA

Routledge is an imprint of the Taylor & Francis Group, an informa business

www.macat.com
info@macat.com

Cataloguing in Publication Data
A catalogue record for this book is available from the British Library.
Library of Congress Cataloguing-in-Publication Data is available upon request.
Cover illustration: Etienne Gilfillan

ISBN 978-1-912302-37-6 (hardback)
ISBN 978-1-912127-57-3 (paperback)
ISBN 978-1-912281-25-1 (e-book)

Notice
The information in this book is designed to orientate readers of the work under analysis,
to elucidate and contextualise its key ideas and themes, and to aid in the development
of critical thinking skills. It is not meant to be used, nor should it be used, as a
substitute for original thinking or in place of original writing or research. References and
notes are provided for informational purposes and their presence does not constitute
endorsement of the information or opinions therein. This book is presented solely for
educational purposes. It is sold on the understanding that the publisher is not engaged
to provide any scholarly advice. The publisher has made every effort to ensure that
this book is accurate and up-to-date, but makes no warranties or representations with
regard to the completeness or reliability of the information it contains. The information
and the opinions provided herein are not guaranteed or warranted to produce particular
results and may not be suitable for students of every ability. The publisher shall not be
liable for any loss, damage or disruption arising from any errors or omissions, or from
the use of this book, including, but not limited to, special, incidental, consequential or
other damages caused, or alleged to have been caused, directly or indirectly, by the
information contained within.

CONTENTS

THE MACAT LIBRARY

The Macat Library is a series of unique academic explorations of seminal works in the humanities and social sciences – books and papers that have had a significant and widely recognised impact on their disciplines. It has been created to serve as much more than just a summary of what lies between the covers of a great book. It illuminates and explores the influences on, ideas of, and impact of that book. Our goal is to offer a learning resource that encourages critical thinking and fosters a better, deeper understanding of important ideas.

Each publication is divided into three Sections: Influences, Ideas, and Impact. Each Section has four Modules. These explore every important facet of the work, and the responses to it.

This Section-Module structure makes a Macat Library book easy to use, but it has another important feature. Because each Macat book is written to the same format, it is possible (and encouraged!) to cross-reference multiple Macat books along the same lines of inquiry or research. This allows the reader to open up interesting interdisciplinary pathways.

To further aid your reading, lists of glossary terms and people mentioned are included at the end of this book (these are indicated by an asterisk [*] throughout) – as well as a list of works cited.

Macat has worked with the University of Cambridge to identify the elements of critical thinking and understand the ways in which six different skills combine to enable effective thinking.
Three allow us to fully understand a problem; three more give us the tools to solve it. Together, these six skills make up the **PACIER** model of critical thinking. They are:

ANALYSIS – understanding how an argument is built
EVALUATION – exploring the strengths and weaknesses of an argument
INTERPRETATION – understanding issues of meaning

CREATIVE THINKING – coming up with new ideas and fresh connections
PROBLEM-SOLVING – producing strong solutions
REASONING – creating strong arguments

To find out more, visit **WWW.MACAT.COM.**

CRITICAL THINKING AND *THE SELFISH GENE*

Primary critical thinking skill: REASONING
Secondary critical thinking skill: INTERPRETATION

Richard Dawkins provides excellent examples of his reasoning and interpretation skills in *The Selfish Gene*. His 1976 book is not a work of original research, but instead a careful explanation of evolution, combined with an argument for a particular interpretation of several aspects of evolution. Since Dawkins is building on other researchers' work and writing for a general audience, the central elements of good reasoning are vital to his book: producing a clear argument and presenting a persuasive case; organising an argument and supporting its conclusions.

In doing this, Dawkins also employs the crucial skill of interpretation: understanding what evidence means; clarifying terms; questioning definitions; giving clear definitions on which to build arguments. The strength of his reasoning and interpretative skills played a key part in the widespread acceptance of his argument for a gene-centred interpretation of natural selection and evolution – and in its history as a bestselling classic of science writing.

ABOUT THE AUTHOR OF THE ORIGINAL WORK

Richard Dawkins is an evolutionary biologist and an outspoken atheist. Born in Kenya in 1941, he moved to Britain as a boy and studied zoology at Oxford University, eventually becoming a lecturer there in 1970. Dawkins made his name as a writer by mixing difficult ideas with an easy style and 2006's best-selling *The God Delusion* turned him into a household name. That same year Dawkins founded The Richard Dawkins Foundation for Science and Reason, calling for the world of science actively to engage in society.

ABOUT THE AUTHORS OF THE ANALYSIS

Dr Nicola Davis studied cell biology at Durham University and received her PhD from the Research Department of Cell and Developmental Biology at University College London.

ABOUT MACAT

GREAT WORKS FOR CRITICAL THINKING

Macat is focused on making the ideas of the world's great thinkers accessible and comprehensible to everybody, everywhere, in ways that promote the development of enhanced critical thinking skills.

It works with leading academics from the world's top universities to produce new analyses that focus on the ideas and the impact of the most influential works ever written across a wide variety of academic disciplines. Each of the works that sit at the heart of its growing library is an enduring example of great thinking. But by setting them in context – and looking at the influences that shaped their authors, as well as the responses they provoked – Macat encourages readers to look at these classics and game-changers with fresh eyes. Readers learn to think, engage and challenge their ideas, rather than simply accepting them.

'Macat offers an amazing first-of-its-kind tool for interdisciplinary learning and research. Its focus on works that transformed their disciplines and its rigorous approach, drawing on the world's leading experts and educational institutions, opens up a world-class education to anyone.'

Andreas Schleicher,
Director for Education and Skills, Organisation for Economic
Co-operation and Development

'Macat is taking on some of the major challenges in university education … They have drawn together a strong team of active academics who are producing teaching materials that are novel in the breadth of their approach.'

Prof Lord Broers,
former Vice-Chancellor of the University of Cambridge

'The Macat vision is exceptionally exciting. It focuses upon new modes of learning which analyse and explain seminal texts which have profoundly influenced world thinking and so social and economic development. It promotes the kind of critical thinking which is essential for any society and economy. This is the learning of the future.'

Rt Hon Charles Clarke, former UK Secretary of State for Education

'The Macat analyses provide immediate access to the critical conversation surrounding the books that have shaped their respective discipline, which will make them an invaluable resource to all of those, students and teachers, working in the field.'

Professor William Tronzo, University of California at San Diego

WAYS IN TO THE TEXT

KEY POINTS

- Richard Dawkins is a British evolutionary biologist who has written enormously popular books about evolution and religion.
- Published in 1976, *The Selfish Gene* argues that evolution operates at the level of genes, rather than at the level of individual "animals" or groups of "animals".
- The book made complex scientific concepts accessible to the general public and has also proved important for both evolutionary biologists and philosophers.

Who Was Richard Dawkins?

Richard Dawkins is an evolutionary biologist, named by respected British current affairs magazine Prospect as Britain's top public intellectual. He is famous for his scientific work and as an outspoken atheist.

Dawkins was born in Kenya in 1941 and moved to Britain as a boy. There he attended Oundle, a school with strong links to the Church of England, where attending services was compulsory. It was here that he says he became an atheist. He went on to study zoology at Oxford University, graduating in 1962. By 1967, Dawkins had begun his academic career teaching at the University of California in Berkeley. He returned to Oxford as a lecturer in 1970.

Dawkins published *The Selfish Gene* in 1976, and followed it with The Extended Phenotype in 1982. He achieved early success by mixing complex scholarly ideas with an easy-to-read style, aiming at the expert, the student, and the layman. This made his books popular with both scientists and the general reading public.

Richard Dawkins became a major public figure with his controversial book The God Delusion in 2006. This best-selling work was forcefully anti-religion, calling religious faith a "fixed false belief." The book endeared him to those interested in "new atheism," a belief system that maintains all religion is bad for humanity.

Dawkins has founded The Richard Dawkins Foundation for Science and Reason, which calls for the world of science actively to engage in society.

What Does *The Selfish Gene* Say?

The Selfish Gene is a book about evolution and more specifically about natural selection.* Natural selection is the way in which certain traits become more common in individual organisms—or "animals"— across the generations. Arctic hares, for example, change from blue-gray to white in the winter because this helps them blend into the snow. Blending into the snow meant that white hares were less likely to be eaten and could procreate, creating more white hares, while darker ones were killed off before they could procreate. Over time, more lighter-colored hares were born until Arctic hares were generally white.

At the time of *The Selfish Gene* there was a debate among scientists about how natural selection worked, with three different theories. Did it operate at the level of:

- groups of organisms?
- individual organisms?
- the genes themselves?

The Selfish Gene argued that natural selection worked at the level of genes.

Put simply, Dawkins was arguing that the gene was king in the process of natural selection and that individual organisms would sometimes sacrifice themselves in order to promote the passing on of the gene. A parent would protect its offspring, even if it meant dying, in order to pass the gene down through the generations. The successful gene can be described as "selfish" because it puts its own needs—the need to be passed down the generations—before the needs of the individual or the group.

This theory of "kin selection"* was not new and it was not Dawkins's idea. What Dawkins achieved with *The Selfish Gene* was to extend existing arguments and put them into a language that most people could understand.

He contributed to furthering the argument by redefining the concept of genes to include their role as "replicators"—something that passes DNA* down the generations. He introduced new concepts such as memes,* which are an idea or a behavior transmitted from one organism—one person—to another, like genes through generations. He also established the idea of the organism as a "vehicle." An animal was, he explained, a robot whose only purpose was to preserve their genes so they could be passed on down the generations. When it came to human beings, he argued that because of our intelligence, we could outthink the process and resist becoming a slave to the all-powerful gene. The human being could—uniquely among organisms—act rationally rather than by instinct.

Why Does *The Selfish Gene* Matter?

The main importance of *The Selfish Gene* lies in the fact it was so popular. It sold millions and was made into a documentary. Even though its key idea was not original, Dawkins brought it to the general public and became associated with it. He is now considered one of the most influential evolutionary biologists and thinkers.

Although many scientists now agree with his reasoning, they were not always in favor of Dawkins's populist approach. By explaining complex scientific ideas in easy-to-understand language and giving genes personalities—saying that genes "gave orders" to animals, for instance—he lost the respect of certain academics. This style was crucial to the book's popularity with the wider public, however.

The kin selection theory that Dawkins wrote about in *The Selfish Gene* became very influential. More recently, though, scientists have generally become more pluralist, in that they believe natural selection is working on more than one level. They believe that as well as genes driving the process of evolution by natural selection, individual organisms and groups of organisms also have their part to play.

The Selfish Gene was not only important for evolutionary biologists. Philosophers of science such as Daniel Dennett, who works on free will, were also influenced by Dawkins's thinking. Evolution is interesting to him because of the idea that if humans are basically robots for passing genes down the generations, how can we have free will? Do we not just have to do what our genes command? Dennett and Dawkins agree that genes guide our behavior, but that humans are rational enough to ignore those genes' demands if we choose.

Another reason for the attention *The Selfish Gene* received was that Dawkins's arguments were used by conservatives to justify their belief in the free market in economics. They argued that if animals are naturally "selfish" then people are right to do whatever promotes their own self-interest. In later editions, Dawkins made it clear that this was a misunderstanding—selfish behavior and selfish genes are not the same thing and that the selfish gene can actually promote unselfish, or altruistic, behavior.

SECTION 1
INFLUENCES

THE AUTHOR AND THE HISTORICAL CONTEXT

KEY POINTS

- Dawkins studied zoology at Oxford University, and adopted the gene-centric view of Nikolaas Tinbergen.*

- Dawkins's atheism* arguably influenced his scientific ideas.

- Sociobiology*—the biological determination of social behavior—was politically controversial because of its connection to the discredited pseudoscientific field of eugenics*—the idea that the genetic quality of the human population could be improved by selective breeding and forced sterilization.

Why Read this Text?

The Selfish Gene by Richard Dawkins brought important concepts and new terms to the evolution debate for both scientists and the general reading public. Building on original research from leading names in the field—evolutionary biologist William D. Hamilton* and geneticist Luigi L. Cavalli-Sforza*—Dawkins developed an argument for a new view of evolution based on genes,* sequences of DNA* that govern hereditary characteristics that pass on from generation to generation.

Most other scientists up to this point had argued that evolution was driven by individual or group traits, the so-called "survival of the fittest" argument, where those individuals or groups of individuals best adapted to their surroundings (faster runners, better camouflaged) survive to pass their genes on down the generations.

But Dawkins believed that this theory couldn't explain certain behaviors. Why, for example, would a parent sacrifice itself to save its

> **❝** A phenomenon such as Dawkins's *The Selfish Gene* can be seen from many points of view and set in many contexts. Its popular success, its influence on generations of students and scholars, and its permeation of the intellectual life of many countries could all be taken as starting points. **❞**
>
> Alan Grafen,* "The Intellectual Contribution of *The Selfish Gene* to Evolutionary Biology," in *Richard Dawkins: How a Scientist Changed the Way We Think.*

children? His argument was that the gene drives the animal to preserve the *gene*, even ahead of preserving itself.

Author's Life

Richard Dawkins was born in Kenya in 1941 to a British agricultural civil servant and his wife. The family returned to England in 1949 and Dawkins was sent to a school, Oundle, with religious affiliations. It was here, as a teenager, that Dawkins turned his back on Christianity to become an atheist. He moved away from the Church and its belief in a world created by God as described in Genesis—creationism*— and became interested in evolution as an explanation for life on Earth. His farmer parents, who believed that nothing exists beyond the natural world, answered his queries scientifically.

Dawkins studied zoology at Balliol College, Oxford University, where he became a neo-Darwinist* and evolutionary biologist—a student of the science of life and how that life slowly mutates across generations. His mentor there was Nikolaas Tinbergen, a Nobel Prize-winning ethologist* and fellow-atheist. Dawkins graduated in 1962 and, having completed his Ph.D., went to teach zoology at Berkeley in California between 1967 and 1969. He returned to Oxford in 1970, this time to teach.

Dawkins is now known as an outspoken atheist and his views on religion are thought to have contributed to his ideas on evolution. His atheism is touched upon in *The Selfish Gene*, published in 1976, but is discussed more fully in *The Blind Watchmaker* (1986) and *The God Delusion* (2006).

Author's Background

In publishing *The Selfish Gene*, Dawkins entered into a controversial debate between prominent academics that had erupted the previous year. The debate was about sociobiology, the study of the social behavior of animals, particularly in relation to evolutionary adaptation.

Acclaimed American biologist Edward O. Wilson* had argued in his 1975 book *Sociobiology: The New Synthesis*[1] that evolution could account for social behavior. This view would start a political row.

The academic Elizabeth Allen published an article later that year called "Against Sociobiology."[2] In it she forcefully argued that the idea that social behavior and position are determined by nature—what she labeled "determinist theories"—was very dangerous. It amounted, she wrote, to "a genetic justification of the *status quo* and of existing privileges for certain groups according to class, race, or sex."

She went on to say that attributing social problems to biology was to use the same logic as eugenics,* a highly controversial "scientific" idea that claimed you could improve the genetic quality of the human race by selective breeding among those with "good" genes and the forced sterilization of those considered to have "bad" genes.

This debate may not have shaped Dawkins's ideas in *The Selfish Gene* directly, but it had an effect on some of the controversy his ideas would generate. On the one side were right-wing political movements that used his idea that selfishness was a "natural" behavior to justify free-market economics, basically an economic version of "survival of the fittest." On the other side were philosophers like Mary Midgley,* who argued Dawkins was providing a biological justification for self-interest as the foundation of our morality—what she called "ethical egoism."*

NOTES

1   Edward O. Wilson, *Sociobiology: The New Synthesis*. (Cambridge: Belknap Press, 2000).

2   Elizabeth Allen et al., "Against *Sociobiology*," *The New York Review of Books*, http://www.nybooks.com/articles/archives/1975/nov/13/against-sociobiology/, accessed August 12, 2014.

ACADEMIC CONTEXT

KEY POINTS

- Evolutionary biology is concerned with determining the effects of evolution in general. Sociobiology* is concerned with understanding social behavior as affected by evolution.

- Sociobiology expands upon the seminal work of Charles Darwin,* applying principles of natural selection* to behavioral rather than physical traits.

- *The Selfish Gene* was relevant to the controversial scientific debates of the period.

The Work In Its Context

The Selfish Gene is a work about evolutionary theory in general and sociobiology in particular. Evolutionary theory sets out to determine how natural selection—the way in which certain traits become more common in individual organisms across the generations—works on species, and to chart the course of this evolution. Sociobiology deals more specifically with how social behavior evolves. It is especially interested in examples where that social behavior seems to go against basic individual survival.

Richard Dawkins initially felt that theories about the biological roots of social behavior underplayed the role of evolution. He was not alone, though, and the 1960s and 1970s saw a growing number of scientists begin to research social behavior from the viewpoint of evolution.

Overview of the Field

Sociobiology is an offshoot of evolutionary biology, the field of study

66 In the process of natural selection, then, any device that can insert a higher proportion of certain genes into subsequent generations will come to characterize the species. One class of such devices promotes prolonged individual survival; another promotes superior mating performance and care of the resulting offspring. As more complex social behavior by the organism is added to the genes' techniques for replicating themselves, altruism* becomes increasingly prevalent and eventually appears in exaggerated forms. This brings us to the central theoretical problem of sociobiology: how can altruism, which by definition reduces personal fitness, possibly evolve by natural selection? 99

Edward O. Wilson, *Sociobiology: The New Synthesis*

founded by the great English biologist Charles Darwin. Darwin's hugely influential work *On the Origin of Species by Means of Natural Selection* (1859) introduced the idea of natural selection, the main way in which evolution works. "It may be said," Darwin wrote of his central idea, "that natural selection is daily and hourly scrutinizing, throughout the world, every variation, even the slightest; rejecting that which is bad, preserving and adding up all that is good."[1] In other words, natural selection is the combined effect of individual strands of DNA* reacting to random changes in environment.

The best illustration of natural selection is the peppered moth in England. The peppered moth was originally a mix of white and black, which helped it camouflage itself against birch trees. At the beginning of the nineteenth century, when the pollution created by the Industrial Revolution* turned those trees black, the white moths were left with no camouflage and were eaten. The only survivors were those moths that were darker and could still go undetected against the newly blackened trees. These reproduced until they outnumbered the whiter moths, meaning that the peppered moth species evolved from white to black by reacting to changes in its environment.

Evolution usually happens over a much longer period of time but in this case, due to the drastic change in the environment, the adaptation was equally drastic. Black is not "better" in any objective way to the original coloring of the moth but it suited the change in circumstances. It's interesting that as pollution in England was drastically reduced, meaning that the trees turned back to white, later generations of moths readapted, turning whiter again.

Sociobiology applies this same principle of adaptation that changed the moths' color to the behaviors of primitive "non-complex" social animals like ants, birds, pack mammals, and early humans. In his controversial 1975 book *Sociobiology: The New Synthesis*, Edward O. Wilson* defined sociobiology as "the systematic study of the biological basis of all social behavior."[2] He added that the study of more complex societies (sociology) needed a more "structuralist* and non-genetic approach."[3]

Other factors that distinguish sociobiology from evolutionary biology are that it looks for specific social behavior that's more complicated than that of two animals just competing for food. Sociobiologists look at cooperation, coordination, hierarchies (the presence of rank and leaders) and regulation, which means that behavior is continuous and passed down the generations.[4] Wilson demonstrated the difference between a "society" of bees and a "population" of bees by emphasizing these traits. Bees in a single hive communicate and cooperate with each other, something they don't do with the bees from another hive, meaning they are two distinct communities.[5]

Academic Influences

The Selfish Gene was published while arguments were raging between pro-sociobiologists (represented by Edward O. Wilson) and anti-sociobiologists (represented by Elizabeth Allen among others). But the debate was not open to all comers. Scientific authors who wanted to take part had to have the backing of an influential academic publishing company such as Oxford University Press, or prestigious scientific journals such as *Nature* or *The Lancet*. These publishing houses were concerned not only with scientific respectability but also with marketability (more so than the journals, whose content was targeted solely at narrow bands of academics). Happily for Dawkins, Oxford University Press was impressed by *The Selfish Gene*, which it identified as an academically relevant work with commercial possibilities. Soraya de Chadarevian,* in her article "*The Selfish Gene* at 30: The Origin and Career of a Book and Its Title," notes that publishers realized the potential popular appeal of such a project right away. "One of the science editors at [Oxford University Press] with a special interest in science books for general audiences, who was entrusted with the project, was enthralled from the beginning" by its combination of rigor and lucidity.[6]

NOTES

1   Charles Darwin, *On the Origin of Species by Means of Natural Selection* (New York: Appleton and Co., 1915), 80.

2   Edward O. Wilson, *Sociobiology: The New Synthesis* (Cambridge: Belknap Press, 2000), 4.

3   Wilson, *Sociobiology*, 4.

4   Wilson, *Sociobiology*, 10.

5   Wilson, *Sociobiology*, 10.

6   Soraya de Chardarevian, "*The Selfish Gene* at 30: The Origin and Career of a Book and Its Title," *Notes & Records of the Royal Society* 61 no. 1 (January 2007): 31–38.

MODULE 3
THE PROBLEM

KEY POINTS

- The contemporary debate at the time the book was written was about the level at which natural selection* occurs: groups, individuals, or genes.
- Group selectionists believed that natural selection favors pro-group behavior while gene (kin) selectionists believed it favors pro-gene behavior.
- Dawkins united and extended the theories of gene/kin selection* and criticized those of group selection.*

Core Question

The problem Richard Dawkins was tackling in *The Selfish Gene* was a red-hot topic in the academic world at the time: at what level does natural selection occur—at the level of the gene,* the individual, or the group?

The reason it was a burning issue was this level has a direct impact on how we look at the process and the mechanics of evolution. This is especially true in the fields of ethology*—or the study of the habits and instincts of animal behavior—and sociobiology,* both of which are at the core of *The Selfish Gene*.

The publication of *The Selfish Gene* followed several works on the subject: *Sociobiology: The New Synthesis* (1975) by Edward O. Wilson,[1] *Adaptation and Natural Selection* (1966) by George C. Williams,[2] *The Naked Ape* (1967) by Desmond Morris,[3] and *The Social Contract* (1970) by Robert Ardrey.[4]

Wilson's *Sociobiology* laid the ground for the success of Dawkins's *The Selfish Gene*. It opened the debate about animal behavior from an evolutionary perspective, but was inaccessible to the general non-

> 66 It is possible to distinguish two rather different processes, both of which could cause the evolution of characteristics which favor the survival, not of the individual, but of other members of the species. These processes I will call kin selection and group selection, respectively. 99
>
> John Maynard Smith, "Group Selection and Kin Selection: A Rejoinder," in *Group Selection*

academic reader even though it was aiming for a wider audience. Dawkins saw that there was a gap between the more readable books on the same subject, which he felt often misrepresented a scientific understanding of evolution, and those that had academic respect but were too dry and technical. He set out to write *The Selfish Gene* to be both readable and enjoyable but also scientifically correct.

The Participants

It was Charles Darwin* who first suggested individual selection as the level of natural selection, and this view was expanded on by British evolutionary biologist William D. Hamilton* in the second half of the twentieth century.

According to this theory, those individuals best suited to their environment will survive and have offspring, passing on the characteristics that assisted in their own survival—the so-called "survival of the fittest." But what about altruistic, or unselfish, behavior that didn't seem to fit into the "survival of the fittest" theory? That was the concern of the relatively new school of sociobiology. The debate centered on whether unselfish behavior could be explained by group selection or kin selection.

Group selection theory states that "characteristics that may be disadvantageous to an individual can persist or increase in the population if they contribute to the survival and reproduction of the group as a whole."[5] British zoologist Vero C. Wynne-Edwards* and American anthropologist Robert Ardrey were the figureheads of the group selection movement. Wynne-Edwards, in his 1962 book *Animal Dispersion in Relation to Social Behavior*,[6] argued that behavioral adaptations occurred within groups in a population, and were selected by survival or extinction of those groups. Richard Dawkins and *The Selfish Gene* were on the opposing team.

The Selfish Gene's argument was that selection occurs at the level of the genes. Kin theory has it that "characteristics that may be disadvantageous to an individual, such as sterility in worker insects or altruistic behavior, can persist or increase in the population if they contribute to the survival and reproduction of the individual's close genetic relatives."[7] In other words, an animal will put the survival of copies of its genes down the generations above its own individual survival or reproduction. Altruism* will be shown if it benefits relatives who carry the same genes. Kin selection was first suggested by Darwin but the concept was popularized among academics by scientific papers written by William D. Hamilton in 1964.

The Contemporary Debate

Dawkins wrote *The Selfish Gene* in conversation with his fellow evolutionary biologists. In it he criticized group selection theorists like Wynne-Edwards and Ardrey and promoted fellow kin selection theorists like William D. Hamilton and George C. Williams for their "insight and visionary quality" in making explicit "the gene's eye view of Darwinism."[8] He wasn't entirely complimentary, though, because he declared that he found their prose too dry.

Dawkins hoped to "help correct the unconscious group-selectionism that then pervaded popular Darwinism."[9] He singled out the writings of Wynne-Edwards and Ardrey and suggested "the apparent existence of individual altruism still has to be explained." He asked whether groups were to be defined as species, family units, mammals, or something even bigger?[10] Dawkins argued that, since groups are often defined as members of the same family or as sharing the same genes, the gene ought to be the level at which selection is examined, as the notion of a group is too arbitrary.

NOTES

1   Edward O. Wilson, *Sociobiology: The New Synthesis* (Cambridge: Belknap Press, 2000).

2   George C. Williams, *Adaptation and Natural Selection: A Critique of Some Current Evolutionary Thought* (Princeton, Princeton University Press, 1966).

3   Desmond Morris, *The Naked Ape: A Zoologist's Study of the Human Animal* (New York: Dell, 1967).

4   Robert Ardrey, *The Social Contract: A Personal Inquiry into the Evolutionary Sources of Order and Disorder* (London: Doubleday, 1970).

5   As defined in the *Oxford English Dictionary* (www.oed.com).

6   Vero C. Wynne-Edwards, *Animal Dispersion in Relation to Social Behavior* (London: Oliver and Boyd, 1962).

7   As defined in the *Oxford English Dictionary* (www.oed.com).

8   Richard Dawkins, *The Selfish Gene* (Oxford: Oxford University Press, 2006), xvi.

9   Dawkins, *Selfish Gene*, xvi.

10 Dawkins, *Selfish Gene*, 10.

THE AUTHOR'S CONTRIBUTION

KEY POINTS

- Dawkins turned kin selectionism* into a "grand" united theory.

- He introduced a number of useful concepts such as "meme,"* which are still in use today.

- Dawkins used the original research of George C. Williams* and William D. Hamilton* as the basis for his study, which contributed to the understanding of the role of genes* in evolution among a vast audience.

Author's Aims

In writing *The Selfish Gene*, Richard Dawkins set out to produce a book that was both scientifically relevant and accessible to the general reader. His work was largely conceptual, which means he thought through the original research of others in order to produce new ideas that are still used today.

Alan Grafen,* Professor of Theoretical Biology at Oxford University, believed *The Selfish Gene*'s enduring contribution was in the power of its interpretation of evolution. He admired the way Dawkins had taken recent works by adaptationists,* such as William D. Hamilton on inclusive fitness,* and had "establish[ed] their unity under Darwinism by interpreting them all in the logical framework of replicators."[1]

Dawkins also took on board theories such as evolutionary stable strategies (ESS)*—strategies that if widespread in a population cannot be driven out by a less popular alternative—and reciprocal altruism,* whereby an organism acts for the benefit of another at its own expense,

> **❝ In the world of the extended phenotype,* ask not how an animal's behavior benefits its genes; ask instead whose genes it is benefiting. ❞**
> Richard Dawkins, *The Selfish Gene*

with the expectation that the beneficiary would do the same at a later date.

These ideas could all be explained with reference to the genes as "immortal" replicators. He described genes as "independent DNA* replicators, skipping like chamois, free and untrammeled down the generations, temporarily brought together in throwaway survival machines, immortal coils shuffling off an endless succession of mortal ones as they forge towards their separate eternities."[2] He was arguing that natural selection* could only be explained by thinking of genes as the units at which evolution acts, including the altruistic behaviors of such organisms as eusocial* (highly social) insects. These behaviors may seem selfless but are actually driven by selfish genes. He called this seemingly improbable theory "stranger than fiction"[3] and was determined to prove it was possible in a gripping way.

Approach

Dawkins did no original research for *The Selfish Gene* and had no intention of breaking new ground in terms of facts. He was looking instead to popularize the ideas he was developing and to extend the work of fellow kin selection theorists. To do this he set out to create new theoretical concepts and ways of describing them. For example, he extended the previous understanding of the gene by referring to it as a "replicator." He separated the genes from the organisms they lived in, which he described as protective "vehicles." Then he made the point that adaptation was there to benefit the gene or replicator only—not the vehicle, or individual.

Dawkins also introduced the idea of memes, which he described as cultural themes and ideas that were "capable of being transmitted from one brain to another."[4] The term "meme" was original, but Dawkins didn't claim the idea as his own. He credited the work of others, like Italian population geneticist Luigi L. Cavalli-Sforza,* with the original idea.

Contribution In Context

Although Dawkins researched ethology* before writing *The Selfish Gene*, the ideas in his book cannot be traced back to his own early academic papers. To track the roots of *The Selfish Gene*, we need to look at the work of William D. Hamilton, one of the first scientists to promote a gene-centered view of evolution, along with the writings of American evolutionary biologist George C. Williams. His 1966 book *Adaptation and Natural Selection*[5] was a great inspiration.

Dawkins acknowledged the influence of a number of evolutionary biologists, but his contribution to the discipline was to look at their findings from an entirely new perspective. His work on evolutionary stable strategies, for example, was influenced by British evolutionary biologist John Maynard Smith* while his writing on reciprocal altruism drew on the work of American sociobiologist* Robert. L. Trivers.*

NOTES

1   Alan Grafen, "The Intellectual Contribution of *The Selfish Gene* to Evolutionary Biology," in *Richard Dawkins: How a Scientist Changed the Way We Think*, ed. Alan Grafen and Mark Ridley (Oxford: Oxford University Press, 2006), 101.

2   Richard Dawkins, *The Selfish Gene* (Oxford: Oxford University Press, 2006), 234.

3   Dawkins, *Selfish Gene*, xxi.

4   Dawkins, *Selfish Gene*.

5   George C. Williams, *Adaptation and Natural Selection: A Critique of Some Current Evolutionary Thought* (Princeton: Princeton University Press, 1966).

SECTION 2
IDEAS

MODULE 5
MAIN IDEAS

KEY POINTS

- The overarching argument in Dawkins's book is that natural selection occurs at the level of replicators (genes*), because only they can trace their existence all the way back through the history of life.

- *The Selfish Gene*'s main theme is the eternal nature of replicators—DNA* molecules should be seen in terms of every copied version of that DNA stretching backward and forward in time, from its existence to its extinction.

- The central idea of *The Selfish Gene* was presented simply, and through metaphor; Dawkins wanted the book to be both accessible to the public and useful to scientists.

Key Themes

Richard Dawkins's central argument in *The Selfish Gene* was that the gene is the only possible basic unit of natural selection,* because genes can copy themselves and be, to all intents and purposes, immortal until all copies die out.

In contrast, neither individuals nor groups of them are directly subject to evolution by means of natural selection. It is the *frequency* of adaptive traits (like a long neck, or white fur) that changes. For example, a giraffe with a long neck that can reach leaves on a tall tree has a fitness advantage over a giraffe with a shorter neck. This means the longer-necked animal will be better at surviving and reproducing, passing its long-neck genes on, than the shorter-necked version, whose short-neck genes will eventually die out.

Dawkins explored how natural selection of genes can account for

> 66 Now they [replicators] swarm in huge colonies,
> safe inside gigantic lumbering robots, sealed off from
> the outside world, communicating with it by tortuous
> indirect routes, manipulating it by remote control. They
> are in you and me; they created us body and mind;
> and their preservation is the ultimate rationale for our
> existence. They have come a long way, those replicators.
> Now they go by the name of genes, and we are their
> survival machines. 99
>
> Richard Dawkins, *The Selfish Gene*

complex social behaviors and adaptations that appear to be hindrances
for individual organisms but which actually serve the gene. Dawkins
discusses a range of both apparently selfish and apparently selfless
behaviors, arguing that, ultimately, they are all governed by the gene
and its selfish need to replicate whatever happens to the "vehicle"
carrying it.

Exploring The Ideas

Dawkins opened his central argument in *The Selfish Gene* with a
discussion of the history of life. He explored the notion of an early
replicator, a simple molecule that could reproduce itself, and described
how copying mistakes could have resulted in the natural selection of
particular replicators with positive attributes. These might be the
capacity to reproduce effectively, longevity or high copying fidelity
(meaning the ability to replicate without error). These replicators, said
Dawkins, now "go by the name of genes."[1]

But unlike those primitive replicators, genes do not exist in
isolation. Dawkins speculated that competition between replicators
for limited resources could have been behind the creation of
"vehicles"—or bodies—where replicators could propagate and be

protected. These vehicles, he argued, would go on to become more and more complex to compete with other replicators until colonies of replicators existed as individual organisms. These organisms were "robot vehicles blindly programmed to preserve […] genes."[2] He called these vehicles "survival machines," claiming that the preservation of our genes is "the ultimate rationale for our existence."[3]

Dawkins was saying that a replicator was basically "immortal:" copies of it stretched back in time and would stretch into the future. "A DNA molecule could theoretically live on in the form of copies of itself for a hundred million years,"[4] he wrote. As the only part of an individual or a species that survives, the gene is a viable candidate for the basic unit of natural selection. As a gene is the survivor long after the individual body has died, it—rather than the body—is the beneficiary of successive adaptations. This was a useful way of looking at evolution, and one that other academics took up and expanded upon.

Language And Expression

Key to *The Selfish Gene* was the language Dawkins used. He wrote to appeal to everyone—the layman, the student, and the expert. He wanted to make difficult and maybe dry topics accessible to people who didn't necessarily have a background in biology, but he didn't want to "dumb down" the ideas. In particular, he hoped that the book might prove to be of "some educational value" to zoology students, as an introduction to difficult topics. He achieved this broad appeal by avoiding scientific language and mathematical formulas, by using metaphors and analogies, and by anthropomorphizing* genes—giving them attributes and personalities usually given only to humans. Dawkins himself acknowledged that analogies can only take you so far, though, and accepted the possibility that "the expert" wouldn't be "totally happy"[5] with his style. Nevertheless, he hoped that *The Selfish Gene* would be beneficial to them, or at least entertaining.[6]

NOTES

1   Richard Dawkins, *The Selfish Gene* (Oxford: Oxford University Press, 2006), 20.

2   Dawkins, *Selfish Gene*, xxi.

3   Dawkins, *Selfish Gene*, 20.

4   Dawkins, *Selfish Gene*, 35.

5   Dawkins, *Selfish Gene*, xxi.

6   Dawkins, *Selfish Gene*, xxii.

MODULE 6
SECONDARY IDEAS

KEY POINTS

- Genes'* effects can be measured by both the behavior of the organism and its effects on the environment.

- The most important secondary idea in *The Selfish Gene* is that of "the extended phenotype," which sees the phenotype* (the expression of genes) as a combination of physical traits and behavioral patterns (including effects on the landscape).

- Dawkins coined the term "meme"* to describe ideas and themes that can be transmitted from one brain to another.

Other Ideas

While the main ideas in *The Selfish Gene* showed how genes should be seen as the basic unit of natural selection,* the secondary ideas extended and complicated these ideas. Secondary ideas centered on how those genes make themselves felt either in behaviors or through effects on the wider environment. Dawkins was looking at the effects of the genes' phenotype: the sum of an individual's observable characteristics, which is regarded as the result of the interaction between that individual's genetic makeup (or genotype) and the world at large.

Exploring The Ideas

Dawkins illustrated his gene-centered view of evolution using a variety of animal behaviors as examples. He referred repeatedly to evolutionary stable strategies (ESS),* a naturally selected but unconsciously adopted tactic that individuals use to compete with

> 66 The argument of this book is that we, and all other animals, are machines created by our genes. Like successful Chicago gangsters, our genes have survived, in some cases for millions of years, in a highly competitive world. This entitles us to expect certain qualities in our genes. I shall argue that a predominant quality to be expected in a successful gene is ruthless selfishness in individual behavior. However, we shall see, there are special circumstances in which a gene can achieve its own selfish goals best by fostering a limited form of altruism* at the level of individual animals. 99
>
> Richard Dawkins, *The Selfish Gene*

each other for a finite resource. Dawkins used the concept to show how the behavior of a group of selfish individuals can make them appear to be one organized unit and suggested that the same logic could be applied to genes, where "a collection of individual selfish entities (genes) can come to resemble a single organized whole."[1]

But the gene's influence is not just on its host "vehicle," it's also on the world at large because of the way that the gene influences the individual's behavior. This he called "the extended phenotype" and used the example of the termite to illustrate his case.

The termite's genes instruct it to build a large mound in which to live and the effect of this large mound on the environment can be seen as a direct result of the termite's genes. This means the mounds are as much a part of the termite's phenotype as how many legs it has. This argument was included in an additional chapter in the second edition of *The Selfish Gene*, which is a short synopsis of Dawkins's second book, *The Extended Phenotype*.[2] He considers the concept his greatest contribution to evolutionary biology.

Dawkins also argues that "cultural transmission is analogous to genetic transmission,"[3] or that language and ideas evolve in the same way that genes and organisms do. He came up with the term "meme" to describe "an entity that is capable of being transmitted from one brain to another"[4] and suggested that an idea passing from one brain to another exhibited the same survival qualities required by replicators for successful genetic natural selection.

Overlooked

Dawkins's argument that the gene is the unit of natural selection is the argument most people focus on in *The Selfish Gene*. What they sometimes overlook is his actual definition of a gene. The accepted definition of a gene is "a sequence of DNA* containing a code for a protein or RNA* [ribonucleic acid, present in all living cells] molecule."[5] Dawkins's definition was wider: to him the term gene meant "any portion of chromosomal material that potentially lasts for enough generations to serve as a unit of natural selection."[6]

For Dawkins, a "gene" was a replicator: a stable, self-replicating molecule. Individual genes cannot be considered replicators, because they don't behave independently during reproduction. And chromosomes* cannot be considered replicators, because they are altered during sexual reproduction. In Dawkins's replicator definition of the gene, he suggested it will "usually be found to lie somewhere on the scale between cistron* and chromosome."[7]

Dawkins's definition of the gene as a replicator has been championed by the likes of David Haig,* a Harvard biologist, in his 2006 essay "The Gene Meme"[8] and is considered helpful in the discussion of the gene-centered view of evolution.

NOTES

1   Richard Dawkins, *The Selfish Gene* (Oxford: Oxford University Press, 2006), 84.

2   Richard Dawkins, *The Extended Phenotype* (Oxford: Oxford University Press, 1982).

3   Dawkins, *Selfish Gene*, 189.

4   Dawkins, *Selfish Gene*, 196.

5   As defined by the *Oxford English Dictionary* (www.oed.com).

6   Dawkins, *Selfish Gene*, 28.

7   Dawkins, *Selfish Gene*, 36.

8   David Haig, "The Gene Meme," in *Richard Dawkins: How a Scientist Changed the Way We Think*, eds. Alan Grafen and Mark Ridley (Oxford: Oxford University Press, 2006), 101.

MODULE 7
ACHIEVEMENT

KEY POINTS

- Dawkins was successful in writing a best-selling book that contributed to scientific debate in evolutionary biology.
- Dawkins's achievement was all the more remarkable for his youth and relative inexperience in scientific debate.
- The title of *The Selfish Gene* made it resonate with right-wing groups as a justification for free-market economics in a way not intended by Dawkins.

Assessing The Argument

One of Richard Dawkins's main intentions in writing *The Selfish Gene* was realized: it had a sizeable impact on public interest in evolutionary biology, animal behavior, and the importance of genetics. It has been republished in second and third editions, as well as being translated into at least 25 languages and has sold more than one million copies since it was first published in 1976. It has also been made into a BBC *Horizon* television documentary.

The Selfish Gene was also highly successful in dismissing theories of group* and individual selection, another of Dawkins's main intentions in writing it. More complex theories have come along that use the selfish gene* theory as their jumping off points, while Dawkins's ideas have become central to modern understanding of evolution. The book celebrated its 30th anniversary with a special extended edition and is still considered highly relevant today.

Achievement In Context

Dawkins's achievement is all the more remarkable considering that he

> **" To Richard Dawkins's own surprise and sometimes alarm ...** *The Selfish Gene* **is now well established as a classic exposition of evolutionary ideas for academic and lay readers alike. "**
>
> Alan Grafen* and Mark Ridley,* "Preface," *Richard Dawkins: How a Scientist Changed the Way We Think*

was a relatively unknown young academic at the time of *The Selfish Gene*. It was his first book and the debate was already dominated by more established scientific writers. Yet Dawkins managed to extend complex ideas while making them useful to other academics and accessible even to the general reader.

Despite his youth there were no significant barriers to his completion of this work, because he had the support of a powerful academic publishing house, which gave his ideas more impact. His writing was slowed down after two chapters by the frequent power cuts experienced in the UK thanks to strike action by miners in the early 1970s. But these events merely delayed rather than severely affected Dawkins's work. He was able to resume his laboratory research and completed the book in 1976.

Limitations

Anyone who can grasp the ideas of evolution and basic genetics can benefit from reading *The Selfish Gene*. Its widely accepted theory of gene-centered evolution is embraced by most academics and now not seen as particularly controversial. What is still controversial is the view that genes are the only unit that natural selection* works on and that individual organisms are merely "vehicles" to protect and propagate selfish genes. Having said that, neither of these theories has ever been proven wrong, and they have not lost their relevance to discussions over the years.

Something Dawkins has had to contend with since the publication of *The Selfish Gene* is misinterpretation, because the book is regarded differently by people from different academic disciplines and different walks of life. Some philosophers criticize it as a justification for egoistical, self-interested behavior, based on their reading of the "selfish" nature of genes. Dawkins has protested that his book is not a moral discussion about the ethics* of humanity, but a book about evolution. However, the book is still often misinterpreted this way. At the same time, champions of free-market economics have promoted the book, using Dawkins's writing to defend economic selfishness (again due to misinterpretation). This is again something Dawkins has argued against.

In the academic world, criticism of his kin selection* theory has recently grown with strong arguments[1] from Harvard professor Martin Nowak* and Edward O. Wilson*[2] leading the debate.

NOTES

1   Roger Highfield, "*The Selfish Gene* is losing friends," *The Telegraph* http://www.telegraph.co.uk/science/evolution/10506006/The-Selfish-Gene-is-losing-friends.html, accessed August 26, 2014.

2   Martin Nowak, Corina Tarnita, and Edward O. Wilson, "The Evolution of Eusociality," *Nature* 466 (2010) 26: 1057.

MODULE 8
PLACE IN THE AUTHOR'S WORK

KEY POINTS

- *The Selfish Gene* is among a minority of Dawkins's books that deal with the science of evolution; the later books are mainly directed against religion.
- Some critics believe Dawkins's early scientific work is more respectable than his recent writings.
- *The Selfish Gene* is Dawkins's most famous book and is among his most widely read.

Positioning

Since *The Selfish Gene* was published in 1976, Richard Dawkins has written 12 books. Among these, only *The Extended Phenotype* continued the gene-centric project that Dawkins started in *The Selfish Gene*. His later books shift towards more general subjects including the relationship between science and the arts, and his promotion of atheism.* But his first book remains one of his most popular, and the term "selfish gene" has become a well-known phrase in the English language.

Starting with *The Blind Watchmaker*, published in 1986, Dawkins started to focus on disproving creationism* more than furthering his arguments on gene-centric evolution. His mission now was to push the message of evolution in general—"to persuade the reader, not just that the Darwinian world-view happens to be true, but that it is the only known theory that could, in principle, solve the mystery of our existence."[1]

This trend in his writing would lead up to *The God Delusion* (2006), where he focused on four key points:

1. You can be an atheist who is happy, balanced, moral, and intellectually fulfilled;

> ❝ Daniel Dennett's* main complaint about my review
> is that I held Dawkins's book to too high a standard.
> *The God Delusion* was, he says, a popular work and, as
> such, one can't expect it to grapple seriously with
> religious thought … [But] the mere fact that a book is
> intended for a broad audience doesn't mean its author
> can ignore the best thinking on the subject … Ironically,
> the clearest evidence comes from Dawkins himself.
> In … *The Selfish Gene*, Dawkins wrestled with the best
> evolutionary thinkers … and presented their ideas in a
> way that could be appreciated by a broad audience. This
> is what made *The Selfish Gene* brilliant; the absence of
> any analogous treatment of religion in Dawkins's new
> book is what makes it considerably less brilliant. ❞
>
> H. Allen Orr,* "Reply to Daniel Dennett," *New York Review of Books*

2. Darwinian natural selection* is a far more credible explanation
of the origin of the biological complexity than creation;

3. There are no religious children, only the children of religious
parents;

4. Non-believers should accept the label of "atheist" proudly.[2]

So, 30 years after *The Selfish Gene*, Dawkins included only one
main point about natural selection in his most famous recent book
and then used it just to argue that it had happened, rather than discuss
how or why it happened.

Integration

The Selfish Gene is probably Dawkins's most influential academic
work, taking "pride of place among his achievements"[3] according to

Alan Grafen* and Mark Ridley* in their 2008 intellectual biography of Dawkins. As Dawkins's focus has shifted from science to philosophy and religious and ethical matters, his work has become less academically important.

While *The Selfish Gene* was championed as a work of scientific seriousness that managed to communicate complex ideas to the general public, critics were less kind about *The God Delusion*, which some saw as less rigorous. In an article reviewing the work for the *New York Review of Books,* H. Allen Orr wrote: "Despite my admiration for much of Dawkins's work, I'm afraid I'm among those scientists who must part company with him here … Indeed, *The God Delusion* seems to me to be badly flawed … For all I know, Dawkins's general conclusion is right. But his book makes a far from convincing case."[4] Orr's opinion was that in Dawkins's later work, controversy and opinion had replaced evidence and academic concerns. "The most disappointing feature of *The God Delusion* is Dawkins's failure to engage religious thought in any serious way . . . [dismissing] simple expressions of faith as base superstition."[5]

Significance

When *The Selfish Gene* was published in 1976 it was welcomed by the academic community but was thought to be "a young man's book," in the words of Arthur Cain,[6] a professor of zoology at the University of Liverpool who had been one of Dawkins's tutors at Oxford University. Cain thought that Dawkins would, over the course of his academic career, change his mind about much of what he argued in *The Selfish Gene*. This was not to be the case, however. In the introduction to the 30th anniversary edition, Dawkins stated that "there is little in the book that I would rush to take back now."[7] In the 1989 edition, he indicated that he valued the book's "youthful quality" and "whiff of revolution"[8] and restricted himself to correcting and commenting on the text in the form of endnotes.

As well as the new endnotes, he also added two new chapters. "The Long Reach of the Gene" was a distillation of *The Extended Phenotype*,[9] his second book, published in 1982. *The Extended Phenotype* contained more original ideas than *The Selfish Gene*, and Dawkins considers this work to be his most important contribution to evolutionary biology.

NOTES

1   Richard Dawkins, *The Blind Watchmaker: Why the Evidence Reveals a Universe Without Design* (New York: Norton, 1987), xiv.

2   Richard Dawkins, *The God Delusion* (London: Bantam Press, 2006) 1–4.

3   Alan Grafen and Mark Ridley, "Preface," *Richard Dawkins: How a Scientist Changed the Way We Think* (Oxford: Oxford University Press, 2008), i.

4   H. Allen Orr, "A Mission to Convert," *New York Review of Books*, http://www.nybooks.com/articles/archives/2007/jan/11/a-mission-to-convert/, accessed August 11, 2014.

5   Orr, "A Mission to Convert."

6   Richard Dawkins, *The Selfish Gene* (Oxford: Oxford University Press, 2006), vii.

7   Dawkins, *Selfish Gene*, vii.

8   Dawkins, *Selfish Gene*, xvii.

9   Richard Dawkins, *The Extended Phenotype* (Oxford: Oxford University Press, 1982).

SECTION 3
IMPACT

THE FIRST RESPONSES

KEY POINTS

- The most strident criticism of the book was that it was unscientific—especially Dawkins's use of popular terms and metaphors rather than equations—and that it provided a biological justification for antisocial and egotistical behaviors.

- Dawkins added endnotes in later editions to add to the scientific credibility of the book, and clarified that he had not set out to provide a biological justification for social behavior.

- Evolutionary biology at the time *The Selfish Gene* was published was extremely focused on the importance of equations to any theory.

Criticism

The Selfish Gene by Richard Dawkins was well received by many but it also attracted a great deal of criticism. Zoologists, sociologists, biologists, and even philosophers attacked many aspects of the book. While some of the criticism was made up of personal attacks on Dawkins's personal credentials or writing style, another significant part was based on his content and scientific accuracy.

A major criticism of the book was Dawkins's reductionist*— overly simplified—viewpoint: the explanation of behavior as nothing more than a consequence of genetic influence. In his review of *The Selfish Gene* in the journal *Nature*, evolutionary biologist and geneticist Richard C. Lewontin* described Dawkins as having fallen into "the old error that all describable behavior must be a direct product of

> ❝ Dawkins is an uncritical philosophic egoist in the first place, and merely feeds the egoist assumption into his *a priori* biological speculations, only rarely glancing at the relevant facts of animal behavior and genetics, and ignoring their failure to support him. There is nothing empirical about Dawkins. ❞
>
> Mary Midgley,* "Gene Juggling," *Philosophy*

natural selection.*"[1]

Dawkins's argument that genes* are the units of natural selection was also criticized from a biological point of view. Harvard evolutionary biologist Stephen Jay Gould* believed strongly, like most other scientists at the time, that natural selection operates at the level of the individual. In his essay "Caring Groups and Selfish Genes"[2] he argued that genes are not directly exposed to natural selection. His argument was that genes do not directly correspond to specific characteristics within an organism, but cooperate with hundreds of other genes to produce any aspect of a body or behavior. He also made the point that evolution by means of natural selection responds to environmental factors that the whole organism experiences. He argued that genes don't respond to these factors, the whole individual does, and so "survival of the fittest" is about individuals not about genes.

Dawkins was also criticized for his redefinition of the gene from the accepted "sequence of DNA* containing a code for a protein or RNA* molecule"[3] to "any portion of chromosomal material that potentially lasts for enough generations to serve as a unit of natural selection"—basically a replicator.[4] Molecular biologist Gunther S. Stent* argued that "this perverse definition denatures the meaningful and well-established central concept of genetics into a fuzzy and heuristically* useless notion."[5] He felt that Dawkins's popularization of the term had gone too far.

Dawkins's use of anthropomorphism*—giving genes personalities when talking about them—brought some of the most fervent criticism even though he made it clear that this was a technique intended to deliver his message more clearly. Statements such as "we have the power to defy the selfish genes of our birth"[6] were called sloppy and misleading.

Criticism came from another direction as well. In a strong critique in 1979 philosopher Mary Midgley attacked Dawkins for what she understood to be a political statement about human behavior.[7] She believed that he was justifying egoism* by suggesting that organisms are genetically selfish. He was also attacked for the idea that organisms were nothing more than "vehicles" for transporting genes down the generations. There were critics among both academics and the general reading public who felt that this denied the possibility of free will.

Author's Response

The fact that the second edition of *The Selfish Gene* was published with few changes in 1989 seems to suggest Dawkins was not taking the criticism on board. He did, however, address his critics in endnotes and in articles. For example, he responded to Mary Midgley's criticism that he was using evolution to make a moral case for selfish behavior by explaining that "genes 'determine' behavior only in a statistical sense."[8] In an earlier article in 1981 called "In Defence of Selfish Genes,"[9] he had addressed her criticisms more fully, informing her that she had misunderstood *The Selfish Gene*: a point other commentators agreed with him on.

He took the argument about *The Selfish Gene* oversimplifying behavior into his next book, *The Extended Phenotype*.[10] Here he explained that he didn't believe in genetic determinism*—the excessive importance given to genes in determining intelligence, behavior, and development—but that sometimes "it is necessary to use language that can be unfortunately misunderstood as genetic determinism."[11] To him the sacrifice of clarity had been necessary at times to reach a wider audience with *The Selfish Gene*.

His disagreement with Stephen Jay Gould dragged on for years—until Gould's death in 2002—with Gould arguing that, rather than gene-level selection being an explanation of evolution, factors such as physical environment, chance, and species extinction needed to be taken into account. Their sustained argument even inspired a book discussing their ideas and opinions, *Dawkins vs. Gould: Survival of the Fittest*.[12]

Conflict And Consensus

In a sense, *The Selfish Gene* was ahead of its time, making claims that mathematics had yet to back up (though this would eventually happen). Dawkins's lack of hard maths would mean skepticism among some scientists.[13] In the endnotes to the second edition of *The Selfish Gene*, he wrote: "One critic complained that my argument was philosophical, as though that was sufficient condemnation."[14]

In a 2006 essay on *The Selfish Gene*, Alan Grafen* addressed the idea that non-mathematical assessments were regarded as less important: "'Where are your equations?' was the not-always-implicit challenge, and those without equations altogether tended to be viewed as time-wasting hopeful simpletons."[15] The lack of mathematic equations in *The Selfish Gene* was one of the reasons it was so accessible to non-scientists, but inevitably one of the reasons he was criticized by some peers. Scientists wanted Dawkins to provide the mathematics to prove his point rather than have "faith" that the mathematics would catch up to his theorizing.[16] Grafen explains that this is the reason *The Selfish Gene* might well have been right, but was nevertheless rejected by certain thinkers at the time.[17]

NOTES

1   Richard C. Lewontin, "Caricature of Darwinism," *Nature* 266 (1976): 283.

2   Stephen Jay Gould, "Caring Groups and Selfish Genes," *Natural History* 86 (1977): 22–24.

3   As defined by the *Oxford English Dictionary* (www.oed.com).

4   Richard Dawkins, *The Selfish Gene* (Oxford: Oxford University Press, 2006), 28.

5   Gunther S. Stent, "You Can Take the Ethics Out of Altruism But You Can't Take the Altruism Out Of Ethics," *Hastings Center Report* (December 1977): 34.

6   Dawkins, *Selfish Gene,* 201.

7   Mary Midgley, "Gene Juggling," *Philosophy* 54 (1979): 439–58.

8   Dawkins, *Selfish Gene*, 267.

9   Richard Dawkins, "In Defence of Selfish Genes," *Philosophy* 56 (1981): 556–73.

10   Richard Dawkins, *The Extended Phenotype* (Oxford: Oxford University Press, 1982).

11   Dawkins, *Extended Phenotype*, 9.

12   Kim Sterelny, *Dawkins vs. Gould: Survival of the Fittest (Revolutions in Science)* (London: Icon Books, 2007).

13   Dawkins, *Selfish Gene*, xxii.

14   Dawkins, *Selfish Gene*, 322.

15   Alan Grafen, "The Intellectual Contribution of *The Selfish Gene* to Evolutionary Biology," in *Richard Dawkins: How a Scientist Changed the Way We Think*, ed. Alan Grafen and Mark Ridley (Oxford: Oxford University Press, 2006), 71.

16   Grafen, "The Intellectual Contribution of *The Selfish Gene*," 71.

17   Grafen, "The Intellectual Contribution of *The Selfish Gene*," 71.

THE EVOLVING DEBATE

KEY POINTS

- Dawkins's gene-centric view of evolution became a kind of new orthodoxy—a generally accepted theory.

- Modern gene-selection theorists have deepened and complicated Dawkins's ideas.

- Evolutionary biologists (Andrew Read,* Helena Cronin*) used *The Selfish Gene* as the basis for a variety of gene-centric theories, while philosophers of science (Daniel Dennett, Kim Sterelny*) were interested in the effects of evolution on free will.

Uses And Problems

The Selfish Gene sparked the debate on evolution among scientists. Unlike many of his peers, Richard Dawkins didn't believe that it was important whether the unit of natural selection* could interact with its environment. To him, what mattered was what "unit" ultimately benefited from the evolution, and for him this was the gene.* He supported this idea by saying that individual organisms can be very different from their parents but the gene itself, still traveling forward down the generations, can remain for the most part unchanged.

The question of what evolution is was at the center of more detailed arguments about whether a unit of natural selection is the same as a unit of evolution, and what it took for a unit to be *the* unit. Dawkins's ideas have been built on and adapted by subsequent scientists, creating alternative theories of evolution. One theory insists that evolution be looked at from the point of view of chemistry—the most fundamental level—another that natural selection operates on

> ❝ So, although we arrive on this planet with a built-in, biologically endorsed set of biases, although we innately prefer certain states of affairs to others, [Dawkins shows] we can nevertheless build lives from this base that overthrow those innate preferences. ❞
>
> Daniel Dennett,* *Elbow Room: The Varieties of Free Will Worth Wanting*

more than one level: genes, individuals, and groups.* This is called the multilevel selection theory.*[1]

The multilevel selection theory has meant the debate about identifying one correct unit of natural selection has cooled down. This new pluralistic perspective was devised by American philosopher Elliott Sober* and evolutionary biologist David S. Wilson* and argues that natural selection acts on several levels at the same time. These pluralists still use the selfish gene theory but have rejected the idea of the gene as the sole unit of natural selection.

Schools Of Thought

The Selfish Gene has been influential among academics in the fields of evolutionary biology and the philosophy of biology and has led to the development of several schools of thought.

One school of thought has developed Dawkins's idea of separating the replicators from the vehicles, the genes from the "host" organism. The school was led by David Hull,* an American philosopher and enthusiastic supporter of Dawkins's ideas on evolution. His contribution to the debate was to assert that the vehicles were not just carriers for genes, but that they had an active role in interacting with the environment. He invented the term "interactors." To him, interactors and replicators both had a part to play in natural selection.[2] Many biologists now consider the interactor to be a crucial part of the unit of selection.

Another school of thought stemming from *The Selfish Gene* was developed by Australian philosopher Kim Sterelny and British philosopher Philip Kitcher.* They supported Dawkins in their 1988 paper "The Return of the Gene,"[3] but put forward a new model, which they called "pluralistic genic selectionism." This term meant that there was no single way to describe the selection process and that targets of selection "do not exist."[4] While they believe that there are many, equally suitable, representations of the evolutionary process, their support for Dawkins comes in the fact that they believe the gene-centric view of natural selection "offers a more general and unified picture of selective processes than can be had from its alternatives."[5]

Dawkins's ideas in *The Selfish Gene* are frequently accused of being reductionist,* but on the positive side these ideas have influenced a reductionist school of thought. In his 2006 book *Darwinian Reductionism: Or, How to Stop Worrying and Love Molecular Biology,*[6] Alexander Rosenberg* argued that natural selection should be viewed at its most fundamental level, that of the physical sciences, particularly chemistry. This school of thought, less popular than the others, argued that different chemical compositions would succeed or fail in different molecular environments. This would mean that these chemical compositions would be more stable and therefore "fitter" and would predominate. Rosenberg argued that this approach allowed a clearer understanding of the lower-level origins of what appeared to be higher-level selections.

In Current Scholarship

The most inevitable followers of *The Selfish Gene* are evolutionary biologists. One of these is Andrew Read of Pennsylvania State University, who researches the evolutionary genetics of pathogens and diseases. In an essay in the collection *Richard Dawkins: How a Scientist Changed the Way We Think,*[7] he described how Dawkins changed the way he thought about evolution and the course of his research. He considers the book a "framework which had tremendous explanatory power for all of biology,"[8] and he now approaches his current research from the gene-centered view of evolution.

Philosophers are also likely to be influenced by Dawkins's text. Helena Cronin, a philosopher of natural and social sciences at the London School of Economics, uses the arguments of *The Selfish Gene* in her work on sexual selection and the differences between the sexes because she believes in its gene-centered approach to natural selection. She has said that *The Selfish Gene* has influenced the direction of her research, because after reading it "*The Selfish Gene* became my staunchest guide."[9] This is a feeling echoed by many academics today, and these researchers continue to interpret and present their findings following Dawkins's framework.

In the field of the philosophy of science, Daniel Dennett and Kim Sterelny have both been particularly interested in Dawkins. Dennett has a specific interest in free will, and in his work takes an evolutionary perspective that is largely in line with Dawkins's beliefs. Kim Sterelny changed the direction of his work as a result of reading Dawkins's theories, moving away from psychology and linguistics and towards evolutionary biology. He explored and developed some of Dawkins's ideas in his paper "The Return of the Gene,"[10] in which he discusses "pluralist gene selectionism," and he has also researched memes,* an idea introduced by Dawkins in *The Selfish Gene*.

NOTES

1   D.S. Wilson and E. Sober, "Reintroducing Group Selection to the Human Behavioral Sciences," *Behavioral and Brain Sciences* 17 (1994): 585–654.

2   David Hull, "Individuality and Selection," *Annual Review of Ecology and Systematics* 11 (1980): 318.

3   Kim Sterelny and Philip Kitcher, "The Return of the Gene," *The Journal of Philosophy* 85 (1988): 339–61.

4   Sterelny and Kitcher, "Return of the Gene," 359.

5   Sterelny and Kitcher, "Return of the Gene," 354.

6   Alexander Rosenberg, *Darwinian Reductionism: Or, How to Stop Worrying and Love Molecular Biology* (Chicago: University of Chicago Press, 2006).

7   Alan Grafen and Mark Ridley, eds., *Richard Dawkins: How a Scientist Changed the Way We Think* (Oxford: Oxford University Press, 2006).

8   Grafen and Ridley, *Richard Dawkins*, 7.

9   Grafen and Ridley, *Richard Dawkins*, 15.

10 Sterelny and Kitcher, "Return of the Gene," 339–61.

IMPACT AND INFLUENCE TODAY

KEY POINTS

- *The Selfish Gene* is used as a teaching tool and represents the basis of more modern theories of gene selection.

- Dawkins's argument with group selection* theorists still rages today, through his negative reviews of Edward O. Wilson's* recent work.

- Wilson and his group selection colleagues respond by criticizing Dawkins directly, suggesting he is a writer about science rather than a scientist.

Position

The Selfish Gene, which provides an excellent introduction to the gene-centered perspective of natural selection,* is currently used as a teaching tool in many universities. Because of the thought-provoking nature of its ideas, it is not limited to the teaching of biology but has also been influential in the philosophy of science.

The long-standing influence of the book is largely linked to Richard Dawkins's fame as a public figure with controversial views on religion, which he frequently airs in the media. People interested in his opinions often start with *The Selfish Gene* because it is his first and best-known book.

The ongoing debate about the value of *The Selfish Gene* has subtly changed as the gene-centered perspective has become increasingly mainstream. Dawkins's ideas have become, in his own words, "more and more the common currency."[1] But that is not to say they are universally accepted. Many academics do not agree that the gene* is the *only* possible unit of selection.

> ❝ Edward Wilson has made important discoveries of his own. His place in history is assured, and so is Hamilton's.* Please do read Wilson's earlier books, including the monumental *The Ants*, written jointly with Bert Hölldobler (yet another world expert who will have no truck with group selection). As for the book under review, the theoretical errors I have explained are important, pervasive, and integral to its thesis in a way that renders it impossible to recommend. To borrow from Dorothy Parker, this is not a book to be tossed lightly aside. It should be thrown with great force. And sincere regret. ❞
>
> Richard Dawkins, reviewing *The Social Conquest of the Earth*, in "The Descent of Edward O. Wilson," *Prospect** magazine

Interaction

The prominent evolutionary biologist Edward O. Wilson, whose *Sociobiology* appeared a year before *The Selfish Gene*, became embroiled in a dispute with Dawkins around his book *The Social Conquest of Earth*. In this 2012 book, Wilson argued against the gene–centric theory of natural selection, stating instead that group selection is behind the evolution of human behavior and altruism.* Among others who reacted to this change of heart from Wilson was Dawkins, who penned a particularly negative review in *Prospect* magazine entitled "The Descent of Edward Wilson." He wrote that the book contained "many pages of erroneous and downright perverse misunderstandings of evolutionary theory,"[2] and drew upon ideas he had presented in *The Selfish Gene* to criticize Wilson. "Wilson was a supporter of [ideas of kin selection], but he has now turned against them in a way that suggests to me that he never really understood them in the first place,"[3] wrote Dawkins. He argued that genes survive through generations and organisms are programmed by genes to help them survive. He went on to declare that in order for groups to be discussed in terms of "inclusive fitness"* their genes must influence the group's development in a way that conveys a phenotype* affecting the whole group's survival and reproduction. "Convincing examples are vanishingly hard to find,"[4] he wrote.

The Continuing Debate

While Wilson's *The Social Conquest of Earth* had its critics, Dawkins obviously included, the *Prospect* magazine review of the book was also criticized. Inevitably, Wilson himself was one of those who attacked the review, responding that the science behind his book had yet to be disproved, especially "by the archaic version of inclusive fitness from the 1970s recited in *Prospect* by Professor Dawkins."[5]

Although most evolutionary scientists have moved on from the kin selection versus group selection debate, Dawkins and Wilson seem determined to continue their personal debate in public. In an interview with the *Guardian* newspaper, Wilson responded to Dawkins's criticism saying, "Dawkins is not a scientist. He's a writer on science and he hasn't participated in research directly or published in peer-reviewed journals for a long time."[6]

NOTES

1   Richard Dawkins, *The Selfish Gene* (Oxford: Oxford University Press, 2006), xv.

2   Dawkins, "The Descent of Edward O. Wilson."

3   Dawkins, "The Descent of Edward O. Wilson."

4   Dawkins, "The Descent of Edward O. Wilson."

5   Edward O. Wilson, reply to Richard Dawkins, "The Descent of Edward O. Wilson."

6   Susanna Rustin, "The Saturday Interview: Harvard Biologist Edward Wilson," *The Guardian*, August 18, 2012.

WHERE NEXT?

KEY POINTS

- *The Selfish Gene* has gone on to inspire a more pluralist view of natural selection,* where multiple levels of selection are considered relevant—genes, individuals, and groups.

- The discussion of the "eusocial gene"*—highly social genes—shows how group-level behaviors can be traced to individual genes.

- Dawkins changed the way scientists communicated with the public, made evolutionary biology a popular subject, and created a unified theory of gene-driven evolution

Potential

Although the evolution debate has moved on to a more multilevel view, *The Selfish Gene* has been a central part of the question of what should be considered the correct unit of natural selection since the 1970s.

Most people working in the field now accept the pluralist view, multilevel selection theory*—where natural selection can operate at different biological levels or even multiple levels at once. Many scientists also agree that different models of natural selection, such as the selfish gene theory or the group selection theory,* can be used together to explain the complex variety of evolutionary dynamics.

Dawkins stands by his belief that the gene-centered view is the *only* correct way to view evolution. And while his perspective is still widely used by academics, the trend has been towards a more pluralistic view, which may affect the impact of *The Selfish Gene* in the future of evolutionary biology.

> **❝ If** *The Selfish Gene* **had not been written when it was, there would still be a need for it to be written today. There are simply no books that have taken its place, even now when so many other books have followed in its wake. ❞**
>
> Marian Dawkins, "Living With *The Selfish Gene*," in *Richard Dawkins: How a Scientist Changed the Way We Think*

The Selfish Gene has formed the basis of mathematical models of evolution such as the one presented in the 2011 article "A Formal Theory of the Selfish Gene,"[1] which backs the gene-selection model with a rigorous theoretical argument. Dawkins's ideas are also continuing to influence not only evolutionary biologists but also philosophers, and even students of English literature. As such, it's likely to continue to be important for many years to come.

Future Directions

Wading into the debate about who is right on the debate between kin* and group* selection (Dawkins and Edward O. Wilson* respectively), Telmo Pievani,* professor of evolutionary biology at the University of Padua, had this to say in 2013: "Both opponents seem to be wrong, facing the general consensus in the field, which favors a pluralistic approach."[2] Having said that, he did give Wilson some credit, accepting that an article Wilson had written in 2010 with Martin Nowak* and Corina Tarnita* had helped push the boundaries of the pluralist approach.

This article had identified a phenomenon call "eusociality, where adult members are divided into reproductive and … non-reproductive castes and the latter care for the young." Wilson, Nowak, and Tarnita went on to ask: "How can genetically prescribed selfless behavior arise by natural selection, which is seemingly its antithesis?"[3] They suggested that there is a "eusocial gene" that is the "center of evolutionary analysis," and that this approach integrates group and kin selection theories. Group behaviors, they argued, can be selected by genes.[4]

Summary

Richard Dawkins's *The Selfish Gene* is important for a number of reasons. First, it has had an impact on how scientists communicate their work to the public. Many consider it to be the first truly accessible biology book, intended for an audience without a background in science, and written in an exciting and gripping style. Previous popular books had still been large and pricey affairs, whereas *The Selfish Gene* was a readable and affordable 200-odd pages. Its popularity had a huge influence on how scientists got their work out to the public. In 1990 Dawkins was even awarded the Michael Faraday Award for his consistent excellence in communicating science to the British public, which really began with *The Selfish Gene*.

Second, *The Selfish Gene* impacted on the public's awareness of evolutionary biology, even introducing terms such as "selfish gene" and "meme"* that are still used today, despite the fact that the book was published in 1976. That said, Dawkins has become such a prominent figure through TV appearances and newspaper articles that his work—including *The Selfish Gene*—has been kept in the public eye.

Third, and most importantly, *The Selfish Gene* had a dramatic impact on the field of evolutionary biology, securing the gene-centered view of evolution its place in the debate. Although Dawkins was re-presenting other scientists' discoveries rather than presenting new research, he did it in such a way as to force a fresh look at the argument that genes are the "immortal" units of natural selection, and the ultimate beneficiaries of adaptive evolutionary change. Although this view has been replaced by a pluralistic one, the ideas in *The Selfish Gene* nevertheless shook up the previously accepted perspectives on the unit of natural selection and changed the way many scientists thought.

NOTES

1   A. Gardner and J.J. Welch, "A Formal Theory of the Selfish Gene," *Journal of Evolutionary Biology* 24 (2011): 1801–13.

2   Telmo Pievani, "Individuals and Groups in Evolution: Darwinian Pluralism and the Multilevel Selection Debate," *Journal of Biosciences* 38 (2013) 4: 1.

3   Martin Nowak et al., "The Evolution of Eusociality," *Nature* 466 (2010) 26: 1057.

4   Nowak et al., "Evolution of Eusociality," 39.

GLOSSARY

GLOSSARY OF TERMS

Adaptationism: a view of evolution that regards the adaptation of an organism to its environment as the principal cause of evolutionary modification; specifically the view that many or most of the characteristics of an organism are adaptations that evolved to fulfill some particular function (*OED*).

Altruism: the practice of doing something in the interest in others even at your own expense.

Anthropomorphism: the ascription of a human attribute or personality to anything impersonal or irrational (*OED*).

Atheism: disbelief in the existence of a "supreme being" or god.

Chromosome: each of the rod-like structures that occur in pairs in the cell nucleus of an animal or plant and hence in every developed cell, and which are carriers of the genes (*OED*).

Cistron: a section of nucleic acid that codes for a specific polypeptide (number of amino acids linked together in a chain, several of which can be linked further to form protein molecules) (*OED*).

Creationism: the belief that the universe was created, along with the Earth and everything in it, according to a religious account. This point of view is often opposed to evolution, and holds that all life was "created" in its present form by God.

DNA (deoxyribonucleic acid): provides the universal blueprint from which all proteins within all organisms are created. It is made up of four different chemicals known as nucleotides (A, T, C and G), and it is the unique combination of these components that determine protein structure and, ultimately, the characteristics of the organism.

Egoism: the theory that regards self-interest as the foundation of morality. Also, in a practical sense: regard to one's own interest, as the supreme guiding principle of action; systematic selfishness. (In recent use opposed to altruism.*) (*OED*).

Ethology: the branch of natural history that deals with the actions and habits of animals, and their reaction to their environment; esp. the study of instinctive animal behavior (*OED*).

Eugenics: a discredited pseudoscientific practice aimed at increasing the genetic quality of the human population through selective breeding and forced sterilization.

Eusocial genes: genes that contribute to a very high level of social organization and cooperation.

Evolutionary Stable Strategy (ESS): a strategy that, if most members of a population adopt it, cannot be bettered by any alternative strategy, as defined in *The Selfish Gene*.

Fecundity: the faculty of reproduction, the capacity for bringing forth young; productiveness (*OED*).

Fidelity: the quality of being faithful. In this context, the copy is faithful to the original, and does not contain mistakes (*OED*).

Gene: described in *The Selfish Gene* as "any portion of chromosomal material (organized and structured DNA* that carries the genes) that potentially lasts for enough generations to serve as a unit of natural selection."

Genetic determinism: the determination of a process or effect by genes; specifically the attribution of sole or excessive importance to genes in the determination of intelligence, behavior, development, etc. (*OED*).

Group selection: the theory of natural selection that holds that characteristics that may be disadvantageous to an individual can persist or increase in the population if they contribute to the survival and reproduction of the group as a whole (*OED*).

Heuristic: allowing you to learn things for yourself in a "hands on" way.

Inclusive fitness: a concept within evolutionary psychology. An organism is judged to have inclusive fitness if it can not only have a large number of offspring, but also support them long enough for them in turn to have offspring, even if it means reducing its own viability.

Industrial Revolution: the period in British history in the late eighteenth and nineteenth centuries that saw an explosion of industrial activity due to the invention of machinery.

Kin selection: a form of natural selection* in which characteristics that may be disadvantageous to an individual, such as sterility in worker insects or altruistic behavior, can persist or increase in the population if they contribute to the survival and reproduction of the individual's close genetic relatives (*OED*).

Meme: a cultural element or behavioral trait whose transmission and consequent persistence in a population, although occurring by non-genetic means (esp. imitation), is considered as analogous to the inheritance of a gene (*OED*).

Multilevel selection: a commonly held pluralist view, which proposes that natural selection* can operate at different biological levels, depending on the particular case, and can even operate at multiple levels at once.

Natural selection: the evolutionary theory, originally proposed by Charles Darwin, of the preferential survival and reproduction of organisms better adapted to their environment (*OED*). The unit-of-selection debate is a discussion about which level of life—the gene, individual, or group—should be considered subject to natural selection.

Neo-Darwinism: a theory of biological evolution (widely accepted since the 1920s) based on Darwin's theory of natural selection* but incorporating the theories of later biologists regarding genes, inheritance, and mutation, particularly those of Weismann and Mendel (*OED*).

Phenotype: the sum total of the observable characteristics of an individual, regarded as the consequence of the interaction of the individual's genotype with the environment; a variety of an organism distinguished by observable characteristics rather than underlying genetic features (*OED*).

Reciprocal altruism: a behavior whereby an organism acts for the benefit of another organism at its own expense, with the expectation that the beneficiary would do the same at a later date.

Reductionism: the practice of describing or explaining a complex (esp. mental, social, or biological) phenomenon in terms of relatively simple or fundamental concepts, especially when this is said to provide a sufficient description or explanation (*OED*).

RNA (ribonucleic acid): a type of nucleic acid present in all living cells and composed of unbranched, often long, chains of ribonucleotides. RNA is principally involved in the synthesis of proteins by transcription and translation of DNA.* RNA differs from DNA in containing the sugar ribose rather than deoxyribose; in having the base uracil in place of thymine; and in usually being single-stranded.

Sociobiology: the study of the social behavior of animals, especially as a means to understanding the biological basis of human social behavior; (in later use) specifically the explanation of social behavior in terms of theories of evolutionary and ecological adaptation (*OED*).

Structuralism: a theoretical idea in the social sciences positing that the elements of social life must be understood in terms of a larger structure of human behavior, rather than the individuals themselves.

PEOPLE MENTIONED IN THE TEXT

Robert Ardrey (1908–80) was an American anthropologist who popularized Vero C. Wynne-Edwards' ideas on the group selection* theory of evolution in his book *The Social Contract* (1970).

Luigi L. Cavalli-Sforza (b. 1922) is an Italian population geneticist, best known for his work on human genetic diversity, who also made contributions to research on language. He summarized his research for the layman in his book *Genes, Peoples, and Languages* (2000).

Arthur Cain (1921–99) was a British evolutionary biologist and fellow of the Royal Academy. He was one of Dawkins's tutors at Oxford University and Professor of Zoology at Liverpool University.

Helena Cronin is a Darwinian philosopher who currently runs the Darwin Centre at the London School of Economics, and is particularly interested in an evolutionary understanding of sex differences. She organized an event in 2006 entitled "The Selfish Gene: Thirty Years On."

Charles Darwin (1809–82) was an English biologist who proposed the theory of evolution through natural selection* in his book *On the Origin of Species* (1859). He is considered one of the most influential figures in our understanding of life.

Marian Dawkins (b. 1945) is a British biologist and professor of animal behavior at Oxford University. She was the first wife of Richard Dawkins.

Soraya de Chadarevian is a professor of history at the University of

California, Los Angeles, specifically in the Center for Society and Genetics.

Daniel Dennett (b. 1942) is an American philosopher of science and of mind. He has a particular interest in evolutionary biology, holding an adaptationist* perspective, in line with Dawkins, that he defended in his book *Darwin's Dangerous Idea* (1995). Along with Dawkins, Sam Harris, and the late Christopher Hitchens, he is known as one of the "Four Horsemen" of modern atheism.*

Stephen Jay Gould (1941–2002) was an American evolutionary biologist and popular science writer. He wrote regular pieces in *Natural History*, and a number of books including *The Panda's Thumb* (1980). He was critical of sociobiology* and criticized Dawkins's ideas.

Alan Grafen is a Scottish professor of theoretical biology in Oxford University's department of zoology.

David Haig (b. 1958) is an Australian evolutionary biologist and professor at Harvard University. He is an expert on the kinship theory of genomic imprinting.

William D. Hamilton (1936–2000) was a British evolutionary biologist who worked on the genetic basis of kin selection and altruism.* His theory is considered by many to be a forerunner of sociobiology.*

David Hull (1935–2010) was an American philosopher interested in the philosophy of biology, particularly evolution. He was an enthusiastic proponent of Dawkins's ideas on evolution, and regularly used his "replicator" terminology.

Philip Kitcher (b. 1947) is a British philosopher who is interested in the philosophy of science, especially of biology. He famously defended Dawkins's selfish gene theory in his 1988 article with Kim Sterelny,* "The Return of the Gene," and has also written a number of books.

Richard C. Lewontin (b. 1929) is an American geneticist and evolutionary biologist at Harvard University. He is a strong opponent of genetic determinism,* including sociobiology,* and he is very critical of many neo-Darwinist* ideas, including those of Richard Dawkins.

John Maynard Smith (1920–2004) was a British evolutionary biologist and geneticist. He formalized the ideas of game theory and evolutionary stable strategies,* which Dawkins drew heavily upon in *The Selfish Gene*.

Mary Midgley (b. 1919) is an English philosopher best known for her work on the philosophy of morals and animal behavior. She wrote a scathing review of *The Selfish Gene*, entitled "Gene Juggling," that led to an ongoing debate.

Desmond Morris (b. 1928) is an English zoologist and ethologist* with a particular interest in sociobiology.*

Martin Nowak (b. 1965) is a professor of evolutionary biology at Harvard University.

H. Allen Orr (b. 1960) is a professor of biology at the University of Rochester.

Telmo Pievani (b.1970) is a professor of evolutionary biology at Padua University in Italy.

Andrew Read (b. 1962) is a New Zealand-born but naturalized British professor of biology and entomology at Pennsylvania State University, an expert on the ecology and evolution of infectious disease.

Mark Ridley (b. 1956) is a British zoologist and writer on evolution currently working at the department of zoology at Oxford University. Dawkins was his doctoral supervisor.

Alexander Rosenberg (b. 1946) is a professor of philosophy at Duke University.

Elliott Sober (b. 1948) is renowned for his work on the philosophy of science, particularly of biology. He helped develop the multilevel selection theory* with David Sloan Wilson* that is discussed in their joint 1999 book *Unto Others: Evolution and Psychology of Unselfish Behavior.*

Gunther S. Stent (1924–2008) was a German molecular biologist at the University of California, Berkeley. He was well known for his writings on the philosophy of biology. He wrote *Molecular Genetics: An Introductory Narrative* (1971).

Kim Sterelny (b. 1950) is an Australian philosopher interested in the philosophy of psychology and biology, particularly of evolution. He famously defended Dawkins's selfish gene theory, along with fellow-philosopher Philip Kitcher,* in their 1988 paper "The Return of the Gene." He has also written about memes* and group selection.*

Corina Tarnita is an assistant professor of biology at Princeton University.

Nikolaas Tinbergen (1907–88) was a Dutch ethologist who won the 1973 Nobel Prize in Physiology or Medicine, for his work on social and individual behavioral patterns in animals. He worked at Oxford University, where he tutored Richard Dawkins.

Robert L. Trivers (b. 1943) is an American evolutionary biologist and sociobiologist.* His key works were on parental investment, parent-offspring conflict and reciprocal altruism,* all key concepts discussed in *The Selfish Gene*.

George C. Williams (1926–2010) was a hugely influential American evolutionary biologist. His first book, *Adaptation and Natural Selection: A Critique of Some Current Evolutionary Thought* (1966), had a major influence on Richard Dawkins when he was writing *The Selfish Gene*.

David S. Wilson (b. 1949) is an American evolutionary biologist. With Elliott Sober,* he co-proposed multilevel selection theory,* a modern version of group selection theory,* in the book *Unto Others: Evolution and Psychology of Unselfish Behavior* (1999). He helped develop and is a prominent advocate of the pluralist multilevel selection theory.*

Edward O. Wilson (b. 1929) is an acclaimed American biologist best known for his research on ants. Like Dawkins, he is a prolific author and has published dozens of books on a range of biological concepts. He is also considered the originator of the field of sociobiology.*

Vero C. Wynne-Edwards (1906–97) was a British zoologist best known for advocating the group selection theory* of evolution, particularly in his book *Animal Dispersion in Relation to Social Behavior* (published in 1962).

WORKS CITED

WORKS CITED

Allen, Elizabeth; Beckwith, Barbara; Beckwith, Jon; Chorover, Steven; Culver, David. "Against *Sociobiology*." *The New York Review of Books*, http://www.nybooks.com/articles/archives/1975/nov/13/against-sociobiology/, accessed August 12, 2014.

Ardrey, Robert. *The Social Contract: A Personal Inquiry into the Evolutionary Sources of Order and Disorder*. London: Doubleday, 1970.

Chardarevian, Soraya de. "*The Selfish Gene* at 30: The Origin and Career of a Book and Its Title." *Notes & Records of the Royal Society* 61 no. 1 (January 2007): 31–38.

Darwin, Charles. *On the Origin of Species by Means of Natural Selection*. New York: Appleton and Co., 1915.

Dawkins, Richard. *The Blind Watchmaker: Why the Evidence Reveals a Universe Without Design*. New York: Norton, 1987.

_____ "In Defence of Selfish Genes." *Philosophy* 56 (1981): 556–73.

_____ "The Descent of Edward O. Wilson." *Prospect* magazine 195 (2012), http://www.prospectmagazine.co.uk/magazine/edward-wilson-social-conquest-earth-evolutionary-errors-origin-species/#.UiyuzsY3uSp, accessed September 8, 2013.

_____ *The Extended Phenotype: The Gene as the Unit of Selection.* Oxford: Oxford University Press, 1982.

_____ *The God Delusion.* London: Bantam Press, 2006.

Dennett, Daniel. *Elbow Room: The Varieties of Free Will Worth Wanting*. Cambridge: MIT Press, 1984.

Gould, Stephen Jay. "Caring Groups and Selfish Genes." *Natural History* 86 (1977): 22–24.

Grafen, Alan, and Mark Ridley, eds. *Richard Dawkins: How a Scientist Changed the Way We Think*. Oxford: Oxford University Press, 2006.

Hull, David. "Individuality and Selection." *Annual Review of Ecology and Systematics* 11 (1980): 311–32.

Lewontin, Richard C. "Caricature of Darwinism." *Nature* 266 (1977): 283–84.

Maynard Smith, John. "Group Selection and Kin Selection: A Rejoinder." In *Group Selection*, ed. George C. Williams, (New Jersey: Transaction Publishers, 1971), 105–113.

Midgley, Mary. "Gene Juggling." *Philosophy* 54 (1979): 439–58.

Morris, Desmond. *The Naked Ape: A Zoologist's Study of the Human Animal.* New York: Dell, 1967.

Nowak, Martin; Tarnita, Corina; and Wilson, Edward O. "The Evolution of Eusocality." *Nature* 466 (2010) 26: 1057–62.

Orr, H. Allen. "A Mission to Convert." *New York Review of Books*, http://www.nybooks.com/articles/archives/2007/jan/11/a-mission-to-convert/, accessed August 11, 2014.

_____ "Reply to Daniel Dennett." *New York Review of Books*, http://www.nybooks.com/articles/archives/2007/mar/01/the-god-delusion/, accessed August 14, 2014.

Pievani, Telmo. "Individuals and Groups in Evolution: Darwinian Pluralism and the Multilevel Selection Debate." *Journal of Biosciences* 38 (2013) 4: 319–25.

Rosenberg, Alexander. *Darwinian Reductionism: Or, How to Stop Worrying and Love Molecular Biology.* Chicago: University of Chicago Press, 2006.

Rustin, Susanna. "The Saturday Interview: Harvard Biologist Edward Wilson." The *Guardian*, August 18, 2012.

Stent, Gunther S. "You Can Take the Ethics Out of Altruism But You Can't Take the Altruism Out Of Ethics." *Hastings Center Report* (December 1977): 33–36.

Sterelny, Kim, and Philip Kitcher. "Return of The Gene." *The Journal of Philosophy* 85 (1988): 339–61.

Sterelny, Kim. *Dawkins vs. Gould: Survival of the Fittest (Revolutions in Science).* London: Icon Books, 2007.

Williams, George C. *Adaptation and Natural Selection: A Critique of Some Current Evolutionary Thought.* Princeton: Princeton University Press, 1966.

Wilson, David S. "Richard Dawkins, Edward O. Wilson, and the Consensus of the Many." *Evolution: This View of Life*, May 29, 2012,

https://evolution-institute.org/article/richard-dawkins-edward-o-wilson-and-the-consensus-of-the-many/ , accessed February 5, 2015.

Wilson, Edward O. *Sociobiology: The New Synthesis* Cambridge: Belknap Press, 2000.

_____ *The Social Conquest of Earth.* New York: Liveright, 2012.

Wyatt Emmerich, J. "Greedy Genes." *The Harvard Crimson*, April 11, 1977.

Wynne-Edwards, Vero C. *Animal Dispersion in Relation to Social Behavior.* Edinburgh and London: Oliver and Boyd, 1962.

THE MACAT LIBRARY
BY DISCIPLINE

AFRICANA STUDIES

Chinua Achebe's *An Image of Africa: Racism in Conrad's Heart of Darkness*
W. E. B. Du Bois's *The Souls of Black Folk*
Zora Neale Huston's *Characteristics of Negro Expression*
Martin Luther King Jr's *Why We Can't Wait*
Toni Morrison's *Playing in the Dark: Whiteness in the American Literary Imagination*

ANTHROPOLOGY

Arjun Appadurai's *Modernity at Large: Cultural Dimensions of Globalisation*
Philippe Ariès's *Centuries of Childhood*
Franz Boas's *Race, Language and Culture*
Kim Chan & Renée Mauborgne's *Blue Ocean Strategy*
Jared Diamond's *Guns, Germs & Steel: the Fate of Human Societies*
Jared Diamond's *Collapse: How Societies Choose to Fail or Survive*
E. E. Evans-Pritchard's *Witchcraft, Oracles and Magic Among the Azande*
James Ferguson's *The Anti-Politics Machine*
Clifford Geertz's *The Interpretation of Cultures*
David Graeber's *Debt: the First 5000 Years*
Karen Ho's *Liquidated: An Ethnography of Wall Street*
Geert Hofstede's *Culture's Consequences: Comparing Values, Behaviors, Institutes and Organizations across Nations*
Claude Lévi-Strauss's *Structural Anthropology*
Jay Macleod's *Ain't No Makin' It: Aspirations and Attainment in a Low-Income Neighborhood*
Saba Mahmood's *The Politics of Piety: The Islamic Revival and the Feminist Subject*
Marcel Mauss's *The Gift*

BUSINESS

Jean Lave & Etienne Wenger's *Situated Learning*
Theodore Levitt's *Marketing Myopia*
Burton G. Malkiel's *A Random Walk Down Wall Street*
Douglas McGregor's *The Human Side of Enterprise*
Michael Porter's *Competitive Strategy: Creating and Sustaining Superior Performance*
John Kotter's *Leading Change*
C. K. Prahalad & Gary Hamel's *The Core Competence of the Corporation*

CRIMINOLOGY

Michelle Alexander's *The New Jim Crow: Mass Incarceration in the Age of Colorblindness*
Michael R. Gottfredson & Travis Hirschi's *A General Theory of Crime*
Richard Herrnstein & Charles A. Murray's *The Bell Curve: Intelligence and Class Structure in American Life*
Elizabeth Loftus's *Eyewitness Testimony*
Jay Macleod's *Ain't No Makin' It: Aspirations and Attainment in a Low-Income Neighborhood*
Philip Zimbardo's *The Lucifer Effect*

ECONOMICS

Janet Abu-Lughod's *Before European Hegemony*
Ha-Joon Chang's *Kicking Away the Ladder*
David Brion Davis's *The Problem of Slavery in the Age of Revolution*
Milton Friedman's *The Role of Monetary Policy*
Milton Friedman's *Capitalism and Freedom*
David Graeber's *Debt: the First 5000 Years*
Friedrich Hayek's *The Road to Serfdom*
Karen Ho's *Liquidated: An Ethnography of Wall Street*

John Maynard Keynes's *The General Theory of Employment, Interest and Money*
Charles P. Kindleberger's *Manias, Panics and Crashes*
Robert Lucas's *Why Doesn't Capital Flow from Rich to Poor Countries?*
Burton G. Malkiel's *A Random Walk Down Wall Street*
Thomas Robert Malthus's *An Essay on the Principle of Population*
Karl Marx's *Capital*
Thomas Piketty's *Capital in the Twenty-First Century*
Amartya Sen's *Development as Freedom*
Adam Smith's *The Wealth of Nations*
Nassim Nicholas Taleb's *The Black Swan: The Impact of the Highly Improbable*
Amos Tversky's & Daniel Kahneman's *Judgment under Uncertainty: Heuristics and Biases*
Mahbub Ul Haq's *Reflections on Human Development*
Max Weber's *The Protestant Ethic and the Spirit of Capitalism*

FEMINISM AND GENDER STUDIES

Judith Butler's *Gender Trouble*
Simone De Beauvoir's *The Second Sex*
Michel Foucault's *History of Sexuality*
Betty Friedan's *The Feminine Mystique*
Saba Mahmood's *The Politics of Piety: The Islamic Revival and the Feminist Subject*
Joan Wallach Scott's *Gender and the Politics of History*
Mary Wollstonecraft's *A Vindication of the Rights of Woman*
Virginia Woolf's *A Room of One's Own*

GEOGRAPHY

The Brundtland Report's *Our Common Future*
Rachel Carson's *Silent Spring*
Charles Darwin's *On the Origin of Species*
James Ferguson's *The Anti-Politics Machine*
Jane Jacobs's *The Death and Life of Great American Cities*
James Lovelock's *Gaia: A New Look at Life on Earth*
Amartya Sen's *Development as Freedom*
Mathis Wackernagel & William Rees's *Our Ecological Footprint*

HISTORY

Janet Abu-Lughod's *Before European Hegemony*
Benedict Anderson's *Imagined Communities*
Bernard Bailyn's *The Ideological Origins of the American Revolution*
Hanna Batatu's *The Old Social Classes And The Revolutionary Movements Of Iraq*
Christopher Browning's *Ordinary Men: Reserve Police Batallion 101 and the Final Solution in Poland*
Edmund Burke's *Reflections on the Revolution in France*
William Cronon's *Nature's Metropolis: Chicago And The Great West*
Alfred W. Crosby's *The Columbian Exchange*
Hamid Dabashi's *Iran: A People Interrupted*
David Brion Davis's *The Problem of Slavery in the Age of Revolution*
Nathalie Zemon Davis's *The Return of Martin Guerre*
Jared Diamond's *Guns, Germs & Steel: the Fate of Human Societies*
Frank Dikotter's *Mao's Great Famine*
John W Dower's *War Without Mercy: Race And Power In The Pacific War*
W. E. B. Du Bois's *The Souls of Black Folk*
Richard J. Evans's *In Defence of History*
Lucien Febvre's *The Problem of Unbelief in the 16th Century*
Sheila Fitzpatrick's *Everyday Stalinism*

The Macat Library By Discipline

Eric Foner's *Reconstruction: America's Unfinished Revolution, 1863-1877*
Michel Foucault's *Discipline and Punish*
Michel Foucault's *History of Sexuality*
Francis Fukuyama's *The End of History and the Last Man*
John Lewis Gaddis's *We Now Know: Rethinking Cold War History*
Ernest Gellner's *Nations and Nationalism*
Eugene Genovese's *Roll, Jordan, Roll: The World the Slaves Made*
Carlo Ginzburg's *The Night Battles*
Daniel Goldhagen's *Hitler's Willing Executioners*
Jack Goldstone's *Revolution and Rebellion in the Early Modern World*
Antonio Gramsci's *The Prison Notebooks*
Alexander Hamilton, John Jay & James Madison's *The Federalist Papers*
Christopher Hill's *The World Turned Upside Down*
Carole Hillenbrand's *The Crusades: Islamic Perspectives*
Thomas Hobbes's *Leviathan*
Eric Hobsbawm's *The Age Of Revolution*
John A. Hobson's *Imperialism: A Study*
Albert Hourani's *History of the Arab Peoples*
Samuel P. Huntington's *The Clash of Civilizations and the Remaking of World Order*
C. L. R. James's *The Black Jacobins*
Tony Judt's *Postwar: A History of Europe Since 1945*
Ernst Kantorowicz's *The King's Two Bodies: A Study in Medieval Political Theology*
Paul Kennedy's *The Rise and Fall of the Great Powers*
Ian Kershaw's *The "Hitler Myth": Image and Reality in the Third Reich*
John Maynard Keynes's *The General Theory of Employment, Interest and Money*
Charles P. Kindleberger's *Manias, Panics and Crashes*
Martin Luther King Jr's *Why We Can't Wait*
Henry Kissinger's *World Order: Reflections on the Character of Nations and the Course of History*
Thomas Kuhn's *The Structure of Scientific Revolutions*
Georges Lefebvre's *The Coming of the French Revolution*
John Locke's *Two Treatises of Government*
Niccolò Machiavelli's *The Prince*
Thomas Robert Malthus's *An Essay on the Principle of Population*
Mahmood Mamdani's *Citizen and Subject: Contemporary Africa And The Legacy Of Late Colonialism*
Karl Marx's *Capital*
Stanley Milgram's *Obedience to Authority*
John Stuart Mill's *On Liberty*
Thomas Paine's *Common Sense*
Thomas Paine's *Rights of Man*
Geoffrey Parker's *Global Crisis: War, Climate Change and Catastrophe in the Seventeenth Century*
Jonathan Riley-Smith's *The First Crusade and the Idea of Crusading*
Jean-Jacques Rousseau's *The Social Contract*
Joan Wallach Scott's *Gender and the Politics of History*
Theda Skocpol's *States and Social Revolutions*
Adam Smith's *The Wealth of Nations*
Timothy Snyder's *Bloodlands: Europe Between Hitler and Stalin*
Sun Tzu's *The Art of War*
Keith Thomas's *Religion and the Decline of Magic*
Thucydides's *The History of the Peloponnesian War*
Frederick Jackson Turner's *The Significance of the Frontier in American History*
Odd Arne Westad's *The Global Cold War: Third World Interventions And The Making Of Our Times*

LITERATURE

Chinua Achebe's *An Image of Africa: Racism in Conrad's Heart of Darkness*
Roland Barthes's *Mythologies*
Homi K. Bhabha's *The Location of Culture*
Judith Butler's *Gender Trouble*
Simone De Beauvoir's *The Second Sex*
Ferdinand De Saussure's *Course in General Linguistics*
T. S. Eliot's *The Sacred Wood: Essays on Poetry and Criticism*
Zora Neale Huston's *Characteristics of Negro Expression*
Toni Morrison's *Playing in the Dark: Whiteness in the American Literary Imagination*
Edward Said's *Orientalism*
Gayatri Chakravorty Spivak's *Can the Subaltern Speak?*
Mary Wollstonecraft's *A Vindication of the Rights of Women*
Virginia Woolf's *A Room of One's Own*

PHILOSOPHY

Elizabeth Anscombe's *Modern Moral Philosophy*
Hannah Arendt's *The Human Condition*
Aristotle's *Metaphysics*
Aristotle's *Nicomachean Ethics*
Edmund Gettier's *Is Justified True Belief Knowledge?*
Georg Wilhelm Friedrich Hegel's *Phenomenology of Spirit*
David Hume's *Dialogues Concerning Natural Religion*
David Hume's *The Enquiry for Human Understanding*
Immanuel Kant's *Religion within the Boundaries of Mere Reason*
Immanuel Kant's *Critique of Pure Reason*
Søren Kierkegaard's *The Sickness Unto Death*
Søren Kierkegaard's *Fear and Trembling*
C. S. Lewis's *The Abolition of Man*
Alasdair MacIntyre's *After Virtue*
Marcus Aurelius's *Meditations*
Friedrich Nietzsche's *On the Genealogy of Morality*
Friedrich Nietzsche's *Beyond Good and Evil*
Plato's *Republic*
Plato's *Symposium*
Jean-Jacques Rousseau's *The Social Contract*
Gilbert Ryle's *The Concept of Mind*
Baruch Spinoza's *Ethics*
Sun Tzu's *The Art of War*
Ludwig Wittgenstein's *Philosophical Investigations*

POLITICS

Benedict Anderson's *Imagined Communities*
Aristotle's *Politics*
Bernard Bailyn's *The Ideological Origins of the American Revolution*
Edmund Burke's *Reflections on the Revolution in France*
John C. Calhoun's *A Disquisition on Government*
Ha-Joon Chang's *Kicking Away the Ladder*
Hamid Dabashi's *Iran: A People Interrupted*
Hamid Dabashi's *Theology of Discontent: The Ideological Foundation of the Islamic Revolution in Iran*
Robert Dahl's *Democracy and its Critics*
Robert Dahl's *Who Governs?*
David Brion Davis's *The Problem of Slavery in the Age of Revolution*

The Macat Library By Discipline

Alexis De Tocqueville's *Democracy in America*
James Ferguson's *The Anti-Politics Machine*
Frank Dikotter's *Mao's Great Famine*
Sheila Fitzpatrick's *Everyday Stalinism*
Eric Foner's *Reconstruction: America's Unfinished Revolution, 1863-1877*
Milton Friedman's *Capitalism and Freedom*
Francis Fukuyama's *The End of History and the Last Man*
John Lewis Gaddis's *We Now Know: Rethinking Cold War History*
Ernest Gellner's *Nations and Nationalism*
David Graeber's *Debt: the First 5000 Years*
Antonio Gramsci's *The Prison Notebooks*
Alexander Hamilton, John Jay & James Madison's *The Federalist Papers*
Friedrich Hayek's *The Road to Serfdom*
Christopher Hill's *The World Turned Upside Down*
Thomas Hobbes's *Leviathan*
John A. Hobson's *Imperialism: A Study*
Samuel P. Huntington's *The Clash of Civilizations and the Remaking of World Order*
Tony Judt's *Postwar: A History of Europe Since 1945*
David C. Kang's *China Rising: Peace, Power and Order in East Asia*
Paul Kennedy's *The Rise and Fall of Great Powers*
Robert Keohane's *After Hegemony*
Martin Luther King Jr.'s *Why We Can't Wait*
Henry Kissinger's *World Order: Reflections on the Character of Nations and the Course of History*
John Locke's *Two Treatises of Government*
Niccolò Machiavelli's *The Prince*
Thomas Robert Malthus's *An Essay on the Principle of Population*
Mahmood Mamdani's *Citizen and Subject: Contemporary Africa And The Legacy Of Late Colonialism*
Karl Marx's *Capital*
John Stuart Mill's *On Liberty*
John Stuart Mill's *Utilitarianism*
Hans Morgenthau's *Politics Among Nations*
Thomas Paine's *Common Sense*
Thomas Paine's *Rights of Man*
Thomas Piketty's *Capital in the Twenty-First Century*
Robert D. Putman's *Bowling Alone*
John Rawls's *Theory of Justice*
Jean-Jacques Rousseau's *The Social Contract*
Theda Skocpol's *States and Social Revolutions*
Adam Smith's *The Wealth of Nations*
Sun Tzu's *The Art of War*
Henry David Thoreau's *Civil Disobedience*
Thucydides's *The History of the Peloponnesian War*
Kenneth Waltz's *Theory of International Politics*
Max Weber's *Politics as a Vocation*
Odd Arne Westad's *The Global Cold War: Third World Interventions And The Making Of Our Times*

POSTCOLONIAL STUDIES

Roland Barthes's *Mythologies*
Frantz Fanon's *Black Skin, White Masks*
Homi K. Bhabha's *The Location of Culture*
Gustavo Gutiérrez's *A Theology of Liberation*
Edward Said's *Orientalism*
Gayatri Chakravorty Spivak's *Can the Subaltern Speak?*

PSYCHOLOGY

Gordon Allport's *The Nature of Prejudice*
Alan Baddeley & Graham Hitch's *Aggression: A Social Learning Analysis*
Albert Bandura's *Aggression: A Social Learning Analysis*
Leon Festinger's *A Theory of Cognitive Dissonance*
Sigmund Freud's *The Interpretation of Dreams*
Betty Friedan's *The Feminine Mystique*
Michael R. Gottfredson & Travis Hirschi's *A General Theory of Crime*
Eric Hoffer's *The True Believer: Thoughts on the Nature of Mass Movements*
William James's *Principles of Psychology*
Elizabeth Loftus's *Eyewitness Testimony*
A. H. Maslow's *A Theory of Human Motivation*
Stanley Milgram's *Obedience to Authority*
Steven Pinker's *The Better Angels of Our Nature*
Oliver Sacks's *The Man Who Mistook His Wife For a Hat*
Richard Thaler & Cass Sunstein's *Nudge: Improving Decisions About Health, Wealth and Happiness*
Amos Tversky's *Judgment under Uncertainty: Heuristics and Biases*
Philip Zimbardo's *The Lucifer Effect*

SCIENCE

Rachel Carson's *Silent Spring*
William Cronon's *Nature's Metropolis: Chicago And The Great West*
Alfred W. Crosby's *The Columbian Exchange*
Charles Darwin's *On the Origin of Species*
Richard Dawkin's *The Selfish Gene*
Thomas Kuhn's *The Structure of Scientific Revolutions*
Geoffrey Parker's *Global Crisis: War, Climate Change and Catastrophe in the Seventeenth Century*
Mathis Wackernagel & William Rees's *Our Ecological Footprint*

SOCIOLOGY

Michelle Alexander's *The New Jim Crow: Mass Incarceration in the Age of Colorblindness*
Gordon Allport's *The Nature of Prejudice*
Albert Bandura's *Aggression: A Social Learning Analysis*
Hanna Batatu's *The Old Social Classes And The Revolutionary Movements Of Iraq*
Ha-Joon Chang's *Kicking Away the Ladder*
W. E. B. Du Bois's *The Souls of Black Folk*
Émile Durkheim's *On Suicide*
Frantz Fanon's *Black Skin, White Masks*
Frantz Fanon's *The Wretched of the Earth*
Eric Foner's *Reconstruction: America's Unfinished Revolution, 1863-1877*
Eugene Genovese's *Roll, Jordan, Roll: The World the Slaves Made*
Jack Goldstone's *Revolution and Rebellion in the Early Modern World*
Antonio Gramsci's *The Prison Notebooks*
Richard Herrnstein & Charles A Murray's *The Bell Curve: Intelligence and Class Structure in American Life*
Eric Hoffer's *The True Believer: Thoughts on the Nature of Mass Movements*
Jane Jacobs's *The Death and Life of Great American Cities*
Robert Lucas's *Why Doesn't Capital Flow from Rich to Poor Countries?*
Jay Macleod's *Ain't No Makin' It: Aspirations and Attainment in a Low Income Neighborhood*
Elaine May's *Homeward Bound: American Families in the Cold War Era*
Douglas McGregor's *The Human Side of Enterprise*
C. Wright Mills's *The Sociological Imagination*

Thomas Piketty's *Capital in the Twenty-First Century*
Robert D. Putman's *Bowling Alone*
David Riesman's *The Lonely Crowd: A Study of the Changing American Character*
Edward Said's *Orientalism*
Joan Wallach Scott's *Gender and the Politics of History*
Theda Skocpol's *States and Social Revolutions*
Max Weber's *The Protestant Ethic and the Spirit of Capitalism*

THEOLOGY

Augustine's *Confessions*
Benedict's *Rule of St Benedict*
Gustavo Gutiérrez's *A Theology of Liberation*
Carole Hillenbrand's *The Crusades: Islamic Perspectives*
David Hume's *Dialogues Concerning Natural Religion*
Immanuel Kant's *Religion within the Boundaries of Mere Reason*
Ernst Kantorowicz's *The King's Two Bodies: A Study in Medieval Political Theology*
Søren Kierkegaard's *The Sickness Unto Death*
C. S. Lewis's *The Abolition of Man*
Saba Mahmood's *The Politics of Piety: The Islamic Revival and the Feminist Subject*
Baruch Spinoza's *Ethics*
Keith Thomas's *Religion and the Decline of Magic*

COMING SOON

Chris Argyris's *The Individual and the Organisation*
Seyla Benhabib's *The Rights of Others*
Walter Benjamin's *The Work Of Art in the Age of Mechanical Reproduction*
John Berger's *Ways of Seeing*
Pierre Bourdieu's *Outline of a Theory of Practice*
Mary Douglas's *Purity and Danger*
Roland Dworkin's *Taking Rights Seriously*
James G. March's *Exploration and Exploitation in Organisational Learning*
Ikujiro Nonaka's *A Dynamic Theory of Organizational Knowledge Creation*
Griselda Pollock's *Vision and Difference*
Amartya Sen's *Inequality Re-Examined*
Susan Sontag's *On Photography*
Yasser Tabbaa's *The Transformation of Islamic Art*
Ludwig von Mises's *Theory of Money and Credit*

Macat Disciplines

Access the greatest ideas and thinkers across entire disciplines, including

Postcolonial Studies

Roland Barthes's *Mythologies*
Frantz Fanon's *Black Skin, White Masks*
Homi K. Bhabha's *The Location of Culture*
Gustavo Gutiérrez's *A Theology of Liberation*
Edward Said's *Orientalism*
Gayatri Chakravorty Spivak's *Can the Subaltern Speak?*

Macat analyses are available from all good bookshops and libraries.

Access hundreds of analyses through one, multimedia tool.
Join free for one month **library.macat.com**

Macat Disciplines

Access the greatest ideas and thinkers across entire disciplines, including

FEMINISM, GENDER AND QUEER STUDIES

Simone De Beauvoir's
The Second Sex

Michel Foucault's
History of Sexuality

Betty Friedan's
The Feminine Mystique

Saba Mahmood's
*The Politics of Piety:
The Islamic Revival and
the Feminist Subject*

Joan Wallach Scott's
*Gender and the
Politics of History*

Mary Wollstonecraft's
*A Vindication of the
Rights of Woman*

Virginia Woolf's
A Room of One's Own

Judith Butler's
Gender Trouble

Macat analyses are available from all good bookshops and libraries.

Access hundreds of analyses through one, multimedia tool.

Join free for one month **library.macat.com**

Macat Disciplines

Access the greatest ideas and thinkers across entire disciplines, including

CRIMINOLOGY

Michelle Alexander's
The New Jim Crow: Mass Incarceration in the Age of Colorblindness

Michael R. Gottfredson & Travis Hirschi's
A General Theory of Crime

Elizabeth Loftus's
Eyewitness Testimony

Richard Herrnstein & Charles A. Murray's
The Bell Curve: Intelligence and Class Structure in American Life

Jay Macleod's
Ain't No Makin' It: Aspirations and Attainment in a Low-Income Neighborhood

Philip Zimbardo's
The Lucifer Effect

Macat analyses are available from all good bookshops and libraries.

Access hundreds of analyses through one, multimedia tool.
Join free for one month **library.macat.com**

Macat Disciplines

Access the greatest ideas and thinkers across entire disciplines, including

Ha-Joon Chang's, *Kicking Away the Ladder*

David Graeber's, *Debt: The First 5000 Years*

Robert E. Lucas's, *Why Doesn't Capital Flow from Rich To Poor Countries?*

Thomas Piketty's, *Capital in the Twenty-First Century*

Amartya Sen's, *Inequality Re-Examined*

Mahbub Ul Haq's, *Reflections on Human Development*

Macat analyses are available from all good bookshops and libraries.

Access hundreds of analyses through one, multimedia tool.
Join free for one month **library.macat.com**

Macat Disciplines

Access the greatest ideas and thinkers across entire disciplines, including

Macat Disciplines

Access the greatest ideas and thinkers across entire disciplines, including

MAN AND THE ENVIRONMENT

The Brundtland Report's, *Our Common Future*
Rachel Carson's, *Silent Spring*
James Lovelock's, *Gaia: A New Look at Life on Earth*
Mathis Wackernagel & William Rees's, *Our Ecological Footprint*

Macat Disciplines

Access the greatest ideas and thinkers across entire disciplines, including

THE FUTURE OF DEMOCRACY

Robert A. Dahl's, *Democracy and Its Critics*
Robert A. Dahl's, *Who Governs?*
Alexis De Toqueville's, *Democracy in America*
Niccolò Machiavelli's, *The Prince*
John Stuart Mill's, *On Liberty*
Robert D. Putnam's, *Bowling Alone*
Jean-Jacques Rousseau's, *The Social Contract*
Henry David Thoreau's, *Civil Disobedience*

Macat Pairs

Analyse historical and modern issues from opposite sides of an argument. Pairs include:

HOW TO RUN AN ECONOMY

John Maynard Keynes's
The General Theory OF Employment, Interest and Money

Classical economics suggests that market economies are self-correcting in times of recession or depression, and tend toward full employment and output. But English economist John Maynard Keynes disagrees.

In his ground-breaking 1936 study *The General Theory*, Keynes argues that traditional economics has misunderstood the causes of unemployment. Employment is not determined by the price of labor; it is directly linked to demand. Keynes believes market economies are by nature unstable, and so require government intervention. Spurred on by the social catastrophe of the Great Depression of the 1930s, he sets out to revolutionize the way the world thinks

Milton Friedman's
The Role of Monetary Policy

Friedman's 1968 paper changed the course of economic theory. In just 17 pages, he demolished existing theory and outlined an effective alternate monetary policy designed to secure 'high employment, stable prices and rapid growth.'

Friedman demonstrated that monetary policy plays a vital role in broader economic stability and argued that economists got their monetary policy wrong in the 1950s and 1960s by misunderstanding the relationship between inflation and unemployment. Previous generations of economists had believed that governments could permanently decrease unemployment by permitting inflation—and vice versa. Friedman's most original contribution was to show that this supposed trade-off is an illusion that only works in the short term.

Macat analyses are available from all good bookshops and libraries.

Access hundreds of analyses through one, multimedia tool.
Join free for one month **library.macat.com**

Macat Pairs

Analyse historical and modern issues from opposite sides of an argument. Pairs include:

ARE WE FUNDAMENTALLY GOOD - OR BAD?

Steven Pinker's
The Better Angels of Our Nature

Stephen Pinker's gloriously optimistic 2011 book argues that, despite humanity's biological tendency toward violence, we are, in fact, less violent today than ever before. To prove his case, Pinker lays out pages of detailed statistical evidence. For him, much of the credit for the decline goes to the eighteenth-century Enlightenment movement, whose ideas of liberty, tolerance, and respect for the value of human life filtered down through society and affected how people thought. That psychological change led to behavioral change—and overall we became more peaceful. Critics countered that humanity could never overcome the biological urge toward violence; others argued that Pinker's statistics were flawed.

Philip Zimbardo's
The Lucifer Effect

Some psychologists believe those who commit cruelty are innately evil. Zimbardo disagrees. In *The Lucifer Effect*, he argues that sometimes good people do evil things simply because of the situations they find themselves in, citing many historical examples to illustrate his point. Zimbardo details his 1971 Stanford prison experiment, where ordinary volunteers playing guards in a mock prison rapidly became abusive. But he also describes the tortures committed by US army personnel in Iraq's Abu Ghraib prison in 2003—and how he himself testified in defence of one of those guards. committed by US army personnel in Iraq's Abu Ghraib prison in 2003—and how he himself testified in defence of one of those guards.

Macat analyses are available from all good bookshops and libraries.

Access hundreds of analyses through one, multimedia tool.
Join free for one month **library.macat.com**

Macat Pairs

*Analyse historical and modern issues
from opposite sides of an argument.
Pairs include:*

Printed in the United States
by Baker & Taylor Publisher Services